A PARADIGM CALLED MAGNETISM

A PARADIGM CALLED MAGNETISM

Sushanta Dattagupta

Indian Institute of Science Education & Research
Kolkata, India

World Scientific

NEW JERSEY • LONDON • SINGAPORE • BEIJING • SHANGHAI • HONG KONG • TAIPEI • CHENNAI

Published by

World Scientific Publishing Co. Pte. Ltd.
5 Toh Tuck Link, Singapore 596224
USA office: 27 Warren Street, Suite 401-402, Hackensack, NJ 07601
UK office: 57 Shelton Street, Covent Garden, London WC2H 9HE

British Library Cataloguing-in-Publication Data
A catalogue record for this book is available from the British Library.

Images on cover design are from
http://www.indiaart.com/Online_Exhibitions/online_exhibitions_1.html

A PARADIGM CALLED MAGNETISM

ISBN-13 978-981-281-386-2
ISBN-10 981-281-386-1

Printed in Singapore.

To the cause of integrated Science Education and Research,
in India.

Preface

Magnetism is truly a very old subject. The first reference to a magnetic material, magnetite or the famed loadstone, can be found in Greek literature in astonishingly early times like 800 B.C. Yet the subject remains modern. It appeals to extremely basic issues of physics–magnetism as a material property is a direct result of one of the fundamental forces of nature governed by electromagnetic interactions. At the same time, this property forms the core of technology, influencing power-generation, biomedical applications through, for instance, magnetic resonance imaging, computer industry with the aid of memory devices and chips, just to name a few. Our objective in this volume is not to dwell on the myriad applications of Magnetism but to highlight the rich structure of its theory. It is remarkable to recount how many extraordinary ideas of physics ensue while teaching a course in Magnetism. To mention some: the first instance of velocity-dependent potentials in Mechanics is encountered in treating the Lorentz force on a charged particle in a magnetic field, the first example of the violation of time reversal is stumbled upon when we consider the Zeeman interaction of a spin in an external magnetic field, and so on.

Given this background to the wide applicability of the concepts of Magnetism, the purpose of these Lecture Notes is to further amplify how the subject has influenced developments in other diverse areas of Physics and how models used in Magnetism can help to clarify a variety of apparently unrelated phenomena. Magnetism is essentially a quantum mechanical subject. Yet its classical limits such as those described by Heisenberg or Ising-like models have far reaching applications to a plethora of topics in phase transitions. Thus, while the ideas of symmetry-breaking and scaling first appeared in Magnetism, they soon pervaded other topics of not just condensed matter physics, but even distant terrains of field theory and high energy physics. Similarly, the concepts of disorder and frustration, embedded in magnetic spin glass systems, are also common to structural glasses. In the domain of non-equilibrium effects too, examples derived from Magnetism help to elucidate the underlying issues of relaxation and dissipation.

With the preceding preamble, the Lecture notes are chapter-wise divided as follows. Chapter 1, consisting of three sections, deals with the by-now well-known phenomena which first appeared in the context of Magnetism but have had important bearing in many other subjects. In section 1.1, we discuss the Mermin-Wagner-Berezinskii theorem on symmetry-breaking in the context of the Heisenberg magnet and its implications in ϕ^4 field theory. The latter forms the backbone of section 1.2 that is devoted to universality and consequent scaling relations which are the underlying concepts in critical point phenomena. In section 1.3 we discuss the issue of multicriticality and point out how the models originally introduced to describe this phenomenon in what are called metamagnets, found their application to tricritical points in 3_{He} - 4_{He} mixtures and even to bicriticality and tetracriticality in an open, nonequilibrium system such as a two-mode ring laser. From chapter 2 onwards, we turn to more recent topics. In chapter 2, we introduce the contemporarily relevant topic of the quantum critical point, again in the context of the Ising model but now in a transverse magnetic field, which finds its realization in rare-earth magnetism. The model helps illustrate the occurrence of quantum phase transitions in a variety of phenomena. The same model allows us to deal with the inter-connected concept of disorder and frustration, which is the subject of chapter 3. Again, the important ideas were first observed in dilute magnets alloyed with metals, called spin glasses. In chapter 4, we move from equilibrium to nonequilibrium statistical mechanics and show how the well-established ideas of spin-lattice and spin-spin relaxations, which are at the heart of magnetic resonance systems, can be further developed to obtain coarse-grained models of phase ordering and pattern formation, seen in many systems as competition between nonlinear interactions and nonequilibrium effects. In chapter 5, we discuss important nonequilibrium phenomena seen in an assembly of single-domain nanomagnetic particles. The concomitant relaxation and memory effects are found to be a consequence of the interplay between polydispersity and inter-particle interactions. Similar memory or aging effects are observed in very different systems of relaxor ferroelectrics, shape-memory ferroelectric materials and structural glasses. The concepts discussed in chapter 4 in the context of magnetic response and relaxation behaviour find their counterparts in the recently-developed subject of dissipative quantum systems, discussed in detail in chapter 6. Therein we analyze the significance of coherence to decoherence transitions using a magnetic paradigm of Diamagnetism, which is intrinsically a quantum phenomenon, for which the boundary of the container plays a critical role. Thus dissipative diamagnetism is prototypical of decoherence in mesoscopic structures. Our final example is from a large spin quantum system that finds its realization in molecular magnets, and in which mesoscopic quantum tunneling can be seen. The influence of environment-induced dissipation makes this system yet another example of observing coherence to decoherence in quantum to classical crossover phenomena.

Acknowledgments

I would like to thank the Department of Science and Technology, New Delhi for awarding me the J.C. Bose Fellowship, which enabled me to complete this project. The staff at World Scientific Press in Singapore deserve special praise for being so patient with me despite me missing so many deadlines for manuscript submission. My wife Ranu, daughters Shahana and Sharmishtha, and numerous colleague island friends have been strongly supportive during my challenging days of transition, when this manuscript was being written. I record my warmest gratitude to them. Finally, I thank Manas Roy for help with the figures and Immanuel Alexander for efficient word-processing.

Sushanta Dattagupta

Contents

Chapter 1

An Assortment of Well-Established Concepts

1.1. Symmetry-Breaking

The dynamics of a many body system are determined in terms of a Hamiltonian. The Hamiltonian is characterized by its invariance properties under all transformations that belong to a group, reflecting the underlying symmetries of the system. An extremely rich example of such a system is the one governed by interacting magnetic spins S and described by what is called the Heinsenberg Hamiltonian:

$$\mathcal{H} = -J \sum_{<ij>} \vec{S}_i \cdot \vec{S}_j . \tag{1.1}$$

The summation is over sites i and j, and is restricted to nearest neighbours, as indicated by the angular brackets $< \;>$. The term J is usually referred to as the exchange interaction [1-5].

We may now enlist all the invariance properties of $\mathcal{H}$. The most basic of all these, as is indeed the fundamental attribute of all mechanical systems, is the invariance under time translation. The implication of this invariance is that the total energy is constant, as it must be for a conservative mechanical system. The second invariance property, again shared by most mechanical systems, is the one exhibited under time reversal. As the time t is reversed to $-t$, each spin operator such as S_i has its sign flipped, because S transforms like the orbital angular momentum operator. There is however no overall change in sign for the pair i and j. Time reversal symmetry is fundamental to electromagnetic forces and is broken only due to dissipation, or the presence of an external magnetic field. The third property, which is what we would like to focus onto, relates to the invariance of $\mathcal{H}$ under the simultaneous rotations of all spins through an arbitrary angle about an arbitrary axis. The group of transformations that leaves the Hamiltonian invariant is the symmetry group, denoted by G.

If the Heisenberg Hamiltonian does not make a distinction between different orientations, as discussed above, how does it then describe ferromagnetism which is associated with a unique directionality of the magnetization? Recall that the thermodynamic state of a system is governed by the minimum of the Helmholtz free energy $F = U - TS$, U being the internal energy, S the entropy and T the temperature. For instance, for the magnetic system at hand, the system would choose those spin configurations for which F is minimized. Thus, at high temperatures, the entropy dominates and the maximally disordered state has the highest entropy. This implies that the high temperature equilibrium state is paramagnetic with no average alignment of spin, i.e. the magnetization is zero. Therefore, the paramagnetic phase is invariant under the same group G as its Hamiltonian. On the other hand, at low temperatures, U dominates over TS and the ground state of $\mathcal{H}$ is the one in which all the spins are aligned along the same direction. (Note the negative sign in the right hand side of Eq. (1).) This low temperature equilibrium phase is a ferromagnetic one with nonzero average spin $<S> = <S_i>$, independent of the site index i, or equivalently a magnetization, $m = \mu V_0^{-1} <S>$, V_0 being the volume of a unit cell and μ the magnetic moment per spin. At a critical temperature T_c called the Curie temperature the system undergoes a phase transition from the entropy dominated paramagnetic state to the energy dominated ferromagnetic state. Thus the magnetization m, which is zero in the paramagnetic phase and becomes non-zero in the ferromagnetic phase, is appropriately called the order parameter of the low temperature phase. It is invariant under rotations about an axis parallel to itself but changes under rotations about all axes that are oblique to itself. Therefore, the ordered phase has a lower symmetry than the full symmetry of the group G — it is a broken symmetry phase, as it breaks the symmetry of the Hamiltonian [6, 7].

The important concept of symmetry-breaking transitions, as enumerated above, transcends the subject of Magnetism to many other areas of Condensed Matter Physics and even to Quantum Field Theory and Particle Physics [8]. Indeed many of the ideas of symmetry and symmetry-breaking transitions in Magnetism, as encapsulated in the Heisenberg or Heisenberg-like models, can be carried over to different fields. We may recall that the most common form of phase transitions in condensed matter physics is encountered in the case of fluids (gases or liquids). A fluid is the highest symmetry phase of matter which is invariant under the Euclidean group of all translations, rotations and reflections. As the external field such as pressure and temperature are varied one sees a series of phase transitions. The most common low temperature phase is a crystalline solid for which the symmetry is lowered from that of the fluid phase. As it turns out the gas-liquid phase transition can be discussed in complete analogy with the magnetic phase transition. While this will be dealt with in Sec. 2 below we want to mention here that in Quantum Field Theory, the idea of symmetry-breaking in a Heisenberg magnet finds its echo in the Lagrangian of n-coupled scalar fields in what is called a ϕ^4 theory. Although the Lagrangian and hence the equations of motion are invariant under a continuous group of transformations, the ground state, called now the vacuum, is not invariant — the symmetry appears spontaneously broken. This phenomenon is completely akin to a Heisenberg magnet which exhibits spontaneous (i.e. in the absence of an applied magnetic field) magnetization below the Curie temperature. Further, just as in the case of Heisenberg magnets wherein low temperature excitations are described by spin waves (whose quanta are magnons), the consequence of continuous symmetry-breaking in field theory is the appearance of massless excitations called the Goldstone Bosons. Indeed, it is this similarity between Magnetism and Particle Physics which has helped establish the deep relationship between Field Theory and Statistical Mechanics [9-11].

To further underscore this similarity, mentioned in the paragraph above, we may recall the famous Mermin-Wagner-Berezinskii (MWB) theorem [12-14]. It states that a system with a continuous symmetry (such as the Heisenberg model under continuous rotations) cannot have a spontaneously broken symmetry in dimensions less than or equal to two. The theorem is most conveniently proved in a simpler version of the

Heisenberg model called the *XY*-model. In the *XY*-model the longitudinal components (along the $\hat{z}$-axis about which the symmetry of the Heisenberg model is expected to be broken) are assumed suppressed. Thus Eq. (1.1) yields

$$\mathcal{H}_{xy} = -J\sum_{<ij>}(S_{ix}S_{jx} \quad S_{iy}S_{jy}). \tag{1.2}$$

The Hamiltonian in Eq. (2), like the one in Eq. (1.1), is invariant under continuous and arbitrary rotations about the $\hat{z}$-axis. In the classical limit Eq. (1.2) may be written as

$$\mathcal{H}_{xy} = -J\sum_{<ij>}\cos(\theta_i - \theta_j). \tag{1.3}$$

Further, in order to obtain the continuous limit of Eq. (1.3) in which the lattice spacing a is arbitrarily small, we can expand the cosine term to yield

$$\mathcal{H} = \frac{J}{4}\sum_{ij}\gamma_{ij}(\theta_i - \theta_j)^2 - \xi JN \quad \text{higher order terms}. \tag{1.4}$$

In Eq. (1.4), ξ is the number of nearest neighbours (the 'coordination number' of the lattice). The restriction on the summations over i and j being over nearest neighbours (indicated by angular brackets in Eq. (1.3)) is removed but is reimposed through the introduction of a matrix γ_{ij} which is defined as

$$\begin{aligned} \gamma_{ij} &= 1, \text{ if } i \text{ and } j \text{ are nearest neighbors}, \\ &= O, \text{ otherwise}. \end{aligned} \tag{1.5}$$

Neglecting the constant and the higher order terms in Eq. (1.4) the continuum version is

$$\mathcal{H}_{xy} = \frac{1}{2}\int d^d x p[\nabla\theta(\vec{x})]^2, \tag{1.6}$$

where d is the spatial dimensionality and

$$p = \frac{\xi J}{2d} a^{2-d} . \tag{1.7}$$

In the *XY* model the spin vector can point in any direction in the *XY*-plane, but if the symmetry is broken, one direction is preferred. Assuming that direction to be the *X*-axis the order parameter is

$$< \cos\theta >= R\left[\frac{1}{Z}\int \mathrm{D}[\theta(\vec{x})]\mathrm{e}^{-\mathrm{i}\theta(\vec{x})}\mathrm{e}^{-\beta\mathcal{H}_{xy}}\right]. \tag{1.8}$$

where Z is the partition function and $\beta = (k_B T)^{-1}$, k_B being the Boltzmann constant and T the temperature. Using the 'Gaussian' property of H_{xy} due to which all cumulants of the fluctuations in $\theta(x)$ beyond the second order vanish, we find

$$< \cos\theta(\vec{x}) > = e^{-\frac{1}{2}<\theta^2(\vec{x})>} . \tag{1.9}$$

Furthermore

$$< \theta^2(\vec{x}) > = -\sum_q \frac{\partial}{\partial\left(\frac{1}{2}\beta\rho q^2\right)} \ln Z_q , \tag{1.10}$$

where $\vec{q}$ is the wave vector that is Fourier-conjugate to $\vec{x}$, and

$$Zq = \left(\frac{2\pi}{\beta\rho q^2}\right)^{\frac{1}{2}} \tag{1.11}$$

Thus

$$< \theta^2(\vec{x}) > = \sum_q \left(\frac{1}{\beta\rho q^2}\right). \tag{1.12}$$

which, in the continuum limit yields

$$< \theta^2(\vec{x}) > = \frac{1}{2\beta\rho} \frac{d^d q}{(2\pi)^d} \frac{1}{q^2} = \frac{1}{4\pi^2 \rho\beta} \frac{\Lambda^{d-2}}{(d-2)} , \tag{1.13}$$

where Λ is the wave number cut off. Equation (1.13) measures the degree to which the order parameter value is diminished from its zero temperature (saturation) value due to classical thermal fluctuations. Evidently as the dimension d approaches the value 2 the fluctuations grow inordinately thereby destroying the order (cf. Eq. (1.9)). This result then proves the celebrated MWB theorem by establishing that fluctuations destroy long range order for a system with continuous symmetry in dimensions less than or equal to two. Physically speaking, the long-wavelength (low q) modes make 'weighty' contributions in low dimensions, thermal excitation of which averages out the orientation of the average spin angular momentum.

The MWB theorem, first derived in the context of magnetism, has been restated by Coleman in Field Theory. As mentioned earlier, spontaneous symmetry-breaking yields a zero mass Goldstone boson. However, in a two-dimensional space-time it is not possible to construct a massless scalar field operator due to severe infrared (or large wavelength) divergences, as indicated above [15].

While the XY-model discussed above is one kind of anisotropy-limit of the Heisenberg model, a different kind of anisotropy ensues when the transverse components of the spin are suppressed and only the longitudinal components are led to interact. This yields the much studied Ising model the Hamiltonian of which can be written as [16, 17]

$$\mathcal{H}_{IS} = -J \sum_{<ij>} S_{iz} S_{jz} \,. \tag{1.14}$$

Unlike the Heisenberg or the XY-model, the Ising model is evidently endowed with discrete symmetry. This symmetry is classified under a group called Z_2 that comprises just two elements: the identity and an element whose square is the identity. This Ising or Z_2 symmetry is broken in those phase transitions which have just two ordered states with order parameters simply differing in sign. Naturally there is no MWB theorem in the Ising case, which does exhibit a phase transition in two dimensions. The discrete symmetry has a further consequence that there is no mass-less Goldstone bosons — the spin wave spectrum (depicting the magnon frequency versus the wave number) has a gap at large wavelengths [18].

The Ising model has a special place in history as it provides a vindication of the subject of statistical mechanics through the path breaking work of Onsager who obtained an exact solution for the partition function of the model in two dimensions. Thus the existence of phase transition, a phenomenon that occupies the centre stage in the subject of thermodynamics, could be demonstrated in terms of a microscopic theory namely that of statistical physics [19].

It is interesting to note that although the Ising model was first introduced as a model of a Heisenberg-like magnet in the limit of extreme anisotropy, such as due to crystal field effects, the majority of its applications is in different fields such as gas-liquid transitions or ordering in alloys. Thus, in the context of gas-liquid transitions, it is a common practice to think of a lattice-gas model wherein a lattice is imagined to consist of cells. Each cell can be occupied by a liquid-like particle, or be empty, to depict a gas-like particle. Hence, the basic statistical variable is a binary one, like in the Ising magnet, but now is a 'pseudo spin' whose upward projection indicates an occupied cell whereas a downward projection implies a vacant cell. If we define the order parameter as the deviation of the density from its average (of the liquid and gas densities of the corresponding coexisting phases) value, the ordered states of liquid and gas are again characterized by equal and opposite (in sign) order parameters, exactly as in the case of the Ising magnet. Indeed this identification of the order parameter allows us to describe the phenomena near the critical point of a gas-liquid phase transition in complete analogy with a paramagnet-to-ferromagnet transition [17]. This is further linked with the concept of 'universality', a subject of Sec. 1.2.

As mentioned above, the discrete binary symmetry of the Ising model finds yet another application to materials science. Consider then a solid alloy of just two elements. Each lattice site is imagined to be occupied by either an A or a B-type of atom. Thus the occupation variable is once again an Ising pseudo-spin, but now one encounters two distinct phenomena depending on whether the exchange parameter J in Eq. (1.14) is positive or negative. If $J > 0$, corresponding to ferromagnetic ordering of the spins in the low-temperature phase, one would see clustering of A-rich (i.e. domains of up spins) and B-rich (i.e. domains of down-spins) phases. The order parameter now is naturally a composition variable, depicting the fractional concentration of one or the other (A or B) type of atoms. Although this analogy between what is called the phase ordering of alloys and the Ising ferromagnet is complete as far as static thermal properties are concerned, the kinetics in the two cases are quite different. If one imagines to 'quench' (i.e. rapidly cool) a spin disordered paramagnet below its Curie temperature, kinetics of ferromagnetic ordering proceed via preferential flipping of spins from one projection to the other. Thus the order parameter, or the magnetization in the present case, is not conserved. On the other hand, in the alloy case, quenching from the disordered (or paramagnetic-like) phase in which the A and B atoms are randomly distributed, to the ordered phase does not imply transmutation of the chemical identify of atoms but simply an interchange of A and B atoms over neighbouring lattice sites. Therefore, the order parameter is now conserved. This difference between conserved and non-conserved order parameter kinetics leads to distinct morphology of domain growth in the two cases [20].

We now discuss the opposite $J < 0$ (antiferromagnetic) case. It is clear from Eq. (1.14) that alternate up and down orientations of the spins lower the energy. Correspondingly, the alloy case leads to alternate ordering of the A and B atoms, reflecting the fact that unlike atoms like each other in contrast to the phase-ordering case wherein like atoms prefer to agglomerate together. The $J < 0$ phase transition is appropriately referred to as the order-disorder transition as can be seen, for example, in beta-brass: a 50-50 admixture of copper and zinc on a BCC lattice. At high temperatures, the Cu and Zn atoms occupy two sub-lattices A and B with equal probability. In the low temperature ordered phase, one set of atoms segregates into the sub-lattice A while the other segregates into the other sub-lattice B. The order parameter: $C = C_{\mathrm{Cu,A}} - C_{\mathrm{Cu,B}}$, where $C = C_{\mathrm{Cu,A}}(C_{\mathrm{Cu,B}})$ is the concentration of Cu atoms in the sub-lattice $A(B)$, is the exact analog of the so-called staggered magnetization of an antiferromagnet. What about kinetics? It is important to note that although kinetic processes involving mutual exchange of A and B atoms on neighbouring sub-lattices conserves the number of A and B atoms, they do not conserve the sub-lattice order parameter.

From the above discussion, it is apparent that a whole range of solidification and ordering phenomena in materials science has derived rich dividends from the analysis of the Ising model in magnetism. In addition to the example of alloys, other transformations that have Ising symmetry are ferroelectric and ferroelastic transitions in which the order parameter is an electric polarization or a spontaneously developed strain, respectively [21]. Further generalization of the Z_2 Ising symmetry to Z_n symmetry yields yet other interesting phenomena of discrete symmetry changes in condensed matter physics. Thus, the $n = 3$ case known as the Potts model [22], leads to surface ordering such as krypton absorbed on graphite in which one finds three equivalent ground states commensurate with one Kr atom for every three unit cells of the graphite lattice [23]. We now have a Z_3 symmetry-broken transition from a fluid to the commensurate state.

1.2. Critical Phenomena and Scaling

Our stated objective in these Lecture Notes has been to employ models and phenomena in Magnetism in order to elucidate developments of similar ideas in other fields. To this end we discuss in this section the extremely important topic of critical phenomena and scaling [24]. Indeed it is not an exaggeration to state that the most profound theoretical advance in physics in the last quarter of the last century has been the successful application of the Renormalization Group (RG) to the critical point phenomena, exploiting and using the ideas of scaling [25-27]. Again it is the second order paramagnetic to ferromagnetic transition at the critical (Curie) point that formed the standard paradigm of critical phenomena, and the Ising and similar models of magnetism became the happy hunting ground for applying RG ideas. We first discuss phase diagrams indicating first order transitions across a two-phase co-existing line culminating at an isolated point of second order transition, called the critical point, and then introduce the idea of scaling.

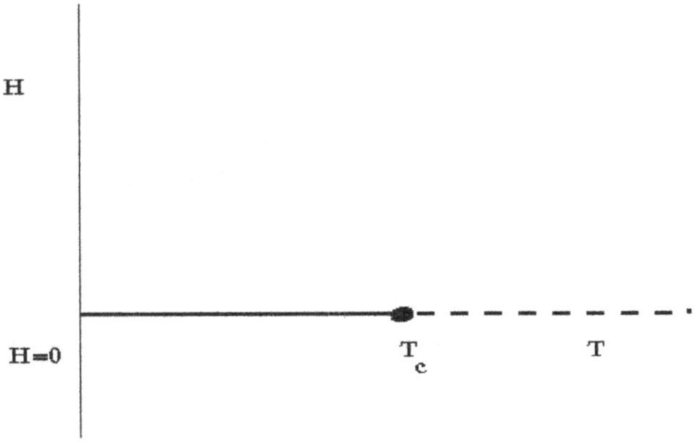

Fig. 1.1. The magnetic field H, applied in the laboratory along the Ising easy axis, is plotted versus the temperature T. The critical point, indicated by a dot, occurs at $H = 0$ and $T = T_c$.

While drawing phase diagrams we follow the Griffiths nomenclature of thermodynamic 'field' variables such as temperature, pressure, magnetic field intensity, etc., which remain the same for co-existing phases, and 'density' variables such as number density, composition, specific volume, magnetization, etc., which acquire distinct values across the co-existence line [28]. The phase diagrams are drawn in either pure field variables space or mixed field-density variables space. An example of the former for a magnet is shown in Fig. 1.1, in which the magnetic field H is plotted versus the temperature T for an Ising model. The co-existence line (for $H = 0$) is shown as a solid line for $T < T_c$ and a dashed line for $T > T_c$; the black dot is the critical point.

Recall from Sec. 1.1 that the symmetry-breaking transition in a magnet occurs spontaneously i.e., in the absence of an external magnetic field. Thus the $H = 0$ line in Fig. 1.1 plays a special role as it is indeed the co-existence line. Thus, experimentally if one had simply gone on to reduce the temperature one would naturally approach T_c from above along the dashed line. Recalling the Z_2 discrete symmetry of a Ising Model (cf. Sec. 1.1) the paramagnetic phase above T_c is a spin-disordered phase for which the macroscopic magnetization M is zero, because it is equally likely to find a spin pointing up or down. The situation is illustrated in Fig. 1.2 which shows the thermodynamic Free Energy F versus M. The curve is schematically a parabola, near $M = 0$ and the thermodynamically stable state for which F is minimum corresponds to $M = 0$. As T approaches and becomes very near to T_c the curvature of F near $M = 0$ becomes vanishingly small (Fig. 1.3) until T is decreased just below T_c when F acquires a double-well structure (Fig. 1.4).

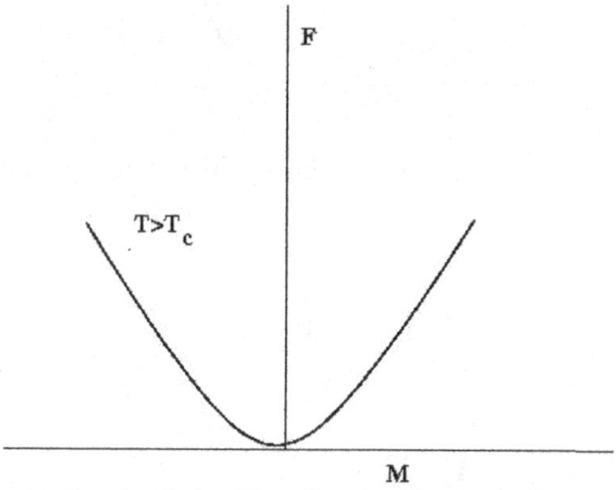

Fig. 1.2. A schematic sketch of the Free Energy F versus the magnetization M, for $T > T_c$.

The two minima of F in Fig. 1.4 are degenerate (i.e. they have the same value of F) corresponding to macroscopic regions in each of which the spins point up or down — the Z_2 symmetry is spontaneously broken. The corresponding situation is illustrated in Fig. 1.5 which depictsthe phase diagram in the mixed density-field variables space, i.e., M versus T. The two minima of F in Fig. 1.4 are denoted by M_+ and M_- ($M_- = -M_+$) which are to be read off from the two ends of what is called a tie line, indicated by a solid vertical line in Fig. 1.5. The length of the tie line goes on increasing as the temperature T goes on decreasing until at $T = 0$ M acquires the saturation value M_s. The entire phenomena, described above, are visualized to occur on the two phase co-existence line (i.e. the solid line in Fig. 1.1). Thus a point on the co-existence line in the pure field space (i.e. in Fig. 1.1), for $T < T_c$ fans out into a line in the mixed space of Fig. 1.5, a feature that is generic and is in conformity with the Gibbs Phase Rule [29].

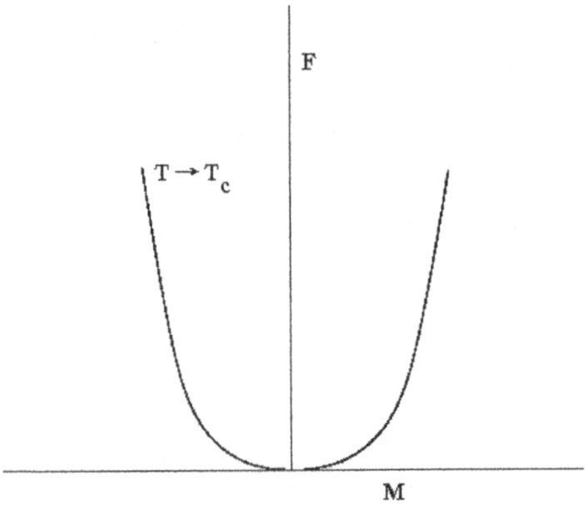

Fig. 1.3. A schematic sketch of the Free Energy F versus the magnetization M, for $T \to T_c$.

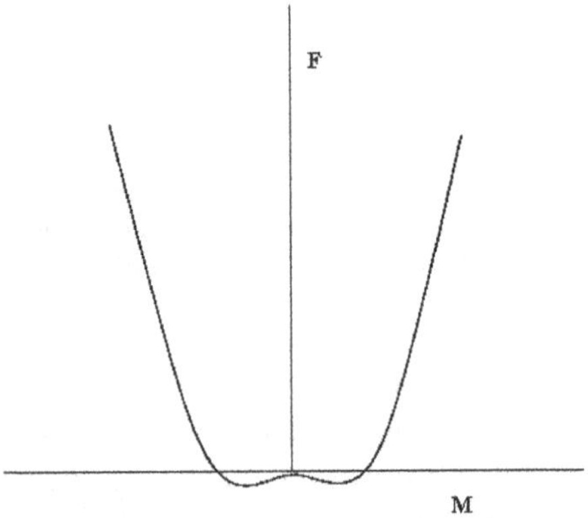

Fig. 1.4. A schematic sketch of the Free Energy F versus the magnetization M, for T just below T_c.

Having discussed the magnetic phase diagram we now analyse in parallel the phase diagram of a fluid that exhibits a gas-liquid phase transition. Recall from Sec. 1.1 that the gas-liquid transition can be described in terms of a lattice gas model which is directly derived from the Ising model. Thus it is not surprising that the two sets of phase diagrams have striking similarties, as well as important dissimilarities. The field that plays the role of the magnetic intensity in a fluid is the pressure P and the conjugate variable, analogous to the magnetization M, is the specific volume V. The P-T phase diagram for a fluid, in the pure field variable space and akin to Fig. 1.1, is shown in Fig. 1.6. Again the solid line (but now a curved one) is a two-phase co-existence line on which the gas and the liquid phases co-exist with the same (degenerate) value of the Free Energy (cf., Fig. 1.4). Further, like in the magnet case, one can go continuously around the critical point (P_c, T_c) from the gas phase to the liquid phase and vice-versa. However, for a fluid, we are more accustomed to think in terms of a first order transition in which the specific volume V jumps discontinuously as we cross the co-existence line, accompanied by a latent heat of vaporization unlike in Fig. 1.5 in which the magnetization grows continuously from its value as T is decreased below T_c. This difference between the two cases is amplified below:

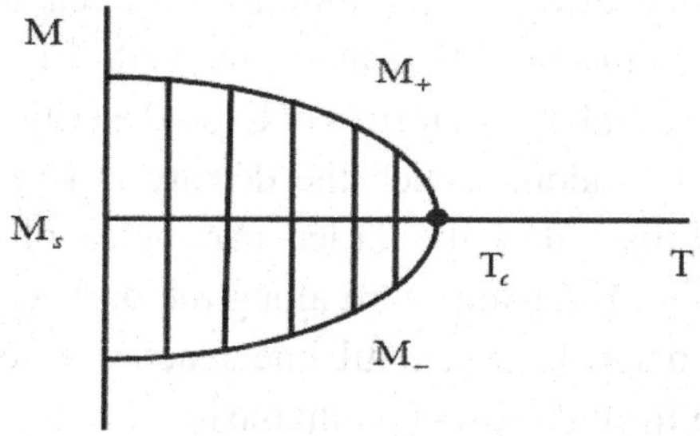

Fig. 1.5 The co-existence line (parabola-shaped) depicting the magnetization M versus the temperature T. The vertical lines are tie-lines, defining the values of M for co-existing phases.

In the magnet case the co-existence line corresponds to $H = 0$, as discussed earlier. This relates to the fundamental symmetry of time reversal invariance, discussed in Sec. 1, which implies that for $H = 0$ the up and down spin projections are equivalent. There is no such symmetry in the fluid case, hence P_c is non-zero and crossing the two-phase co-existence line (the solid one in Fig. 1.6) along an isobar (i.e. constant pressure) or along an isotherm (i.e. constant temperature) would lead to a discontinuous jump of the specific volume.

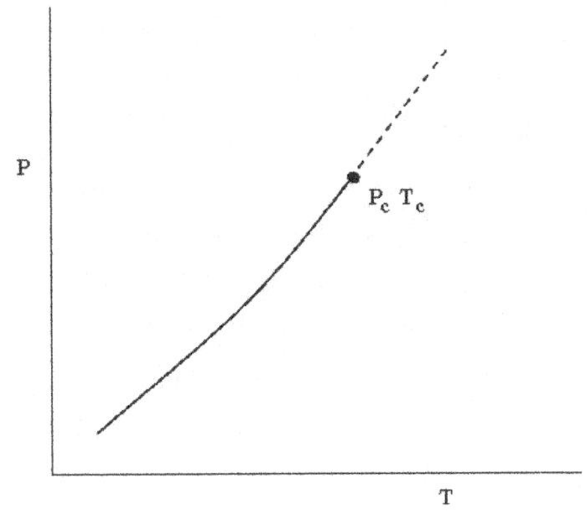

Fig. 1.6. The pressure P versus temperature T for a fluid. The solid line is the gas-liquid co-existence line, the continuation of which (the dashed line) is the critical isochore.

How can then one establish a similarity between the magnet and the fluid cases? The clue is to find the right 'path' of approach to the critical point. That particular path turns out to be the critical isochore (the dashed line in Fig. 1.6) along which the density is fixed at its critical value, say n_c. Defining then the order parameter of the gas-liquid transition as $\phi = n - n_c$, ϕ remains zero along the dashed line, exactly as in Fig. 1.1 for the magnet case, until one reaches below the 'critical point' (P_c, T_c) where from ϕ grows continuously.

The critical point, an isolated point of a second order phase transition as discussed above, is a very special point in the phase diagram of a system. The special nature of the point is related to the fact that certain thermodynamic response functions show divergent behaviour as a function of $t = (T - T_c)/T_c$, as the temperature T approaches the critical temperature T_c from above. Again, the most widely studied response function turns out to be the one relevant for magnetism and that is the magnetic susceptibility χ, which measures the linear response of the magnetization to an applied magnetic field. The function χ diverges as $t^{-\gamma}$, as $t \rightarrow 0^+$ where γ is referred to as a critical exponent. What is further remarkable is not only that a very different property, namely the isothermal compressibility for a different system, i.e. a fluid, is characterized by the same divergent behaviour but that the value of γ is identical to that of an Ising magnet. This follows from the mapping of the Ising model to a lattice-gas model, as mentioned above. The result is quite astonishing because the two systems under consideration are governed by very different interactions and very distinct physical properties. Indeed it is not just a fluid that shows this commonality of behaviour of its response function near the critical point with a magnet but so do other different physical systems e.g. a ferroelectric or an order-disorder alloy, all of which are said to belong to the so-called Ising universality class. This brings us to the issue of universality which links a whole lot of physically different systems based on the common factors of spatial dimensionality d, the dimension n of the order parameter (=1 for the Ising case) and the range of interactions. The astounding statement of universality is that critical exponents like γ (or β which measures how the order parameter such as the magnetization vanishes continuously as $T \rightarrow T_c$) are universal. Thus, to the extent that critical phenomena at a gas-liquid transition in a fluid and a paramagnet-ferromagnet transition in a magnet are describable in terms of the same Ising model (with Z_2 symmetry and short-ranged interactions) in three dimensions, the two very different systems are characterized by the same critical exponents [30].

The universality follows from a very general thermodynamic principle that potentials such as the free energy, are generalized homogeneous functions of temperature and other fields in the vicinity of the critical point. The homogeneity leads to certain scale invariance, the essential idea of which can be summarized as follows.

The occurrence of a critical point is in some sense the culmination of the battle between the energy, which dictates order and the entropy, which embodies disorder. Thus, fluctuations, measured by the correlation function of spins at different sites of an Ising magnet, for instance, grow inordinately. Concomitantly the correlation length, that characterizes the distance over which a pair of spins is correlated, diverges at T_c. Thus all other length scales become irrelevant and the only length scale that matters near T_c is the correlation length. This then is the notion of scaling which allows us to estimate how thermodynamic functions and correlations vary with temperature from the dependence of the correlation length on temperature.

It is this twin concept of universality and scaling that led to the successful application of the renormalization group (RG) by Wilson and others to the critical point phenomena [31]. The RG provides a computational scheme for a class of problems for which fluctuations at multitudes of length scales become important. This scheme was successfully applied by Wilson in the early 1970s to the calculation of critical exponents that led to the theoretical validation of universality. Wilson's original calculation was in the context of field theories in which one introduces a local order parameter, treated as a continuous classical field and the partition function is written as an integral over all possible values of the local order parameter at all space points. But, the most illuminating treatment of RG theory dates back to 1966 when Kadanoff provided a heuristic explanation of scaling, again in the context of an Ising magnet. This real space treatment called Kadanoff construction, does in fact yield a platform on which to build Wilson's RG which allows for explicit calculations of critical exponents and scaling relations. The Kadanoff construction can be exactly carried out for the one-dimensional Isuing model and leads to higher dimensional generalization following an approximate method called the Migdal (1975)- Kadanoff (1976) procedure [32]. A natural extension of the procedure, first developed for an Ising model in $1 + \varepsilon$ dimensions, can be nicely extended to the problem of Krypton adsorbed on graphite which, as mentioned at the end of Chapter 1.1, can be described by the Potts model of Z_3 symmetry [22].

1.3. Multicriticality

The phase diagram for an Ising magnet, discussed in Sec. 1.2 (cf. Fig. 1.1), consists of a single line (for $H = 0$) along the T-axis. The temperature T is thus the only disordering field. Increase in T causes a first order line of two-phase co-existence to terminate at a point called the critical point which is an isolated point of second order transition. What happens when there is more than one disordering field? In the sense of the Gibbs Phase Rule we will have more "degrees of freedom", in which case, is it possible that we will have a line of critical points in a higher dimensional phase diagram? Can we generate special critical points which can be reached by simultaneously fixing two or more disordering fields? The answer is YES, and such critical points, called multi-critical points are the subject of this section.

The present analysis is based on what is now called the Blume-Capel model [33, 34]. Imagine a Hamiltonian which describes an Ising-type interaction between nearest neighbours which however are now spin-1 ions in a lattice that are subject also to a crystal field Δ:

$$\mathcal{H} = -J\sum_{<ij>} S_{ij}S_{jz} + \Delta\sum_{i} S_{iz}^2, \ \Delta > 0 \tag{1.15}$$

Clearly, the crystal field terms would simply be a constant for the ordinary spin half Ising model discussed hitherto. Here $S_i = \pm 1$ or 0 and one now can consider two disordering fields T and Δ shown in Fig. 1.7 For $\Delta = 0$, we are back to our earlier Ising case, in which we have the usual ferromagnetic to paramagnetic transition but now occurring along the ordinate (in constrast to Fig. 1.1). For small enough Δ the single ion energy levels $S_i = \pm 1$ lie not much higher than $S_i = 0$ singlet and so the exchange interaction remains operative. Under this condition, mean field calculations suggest that the paramagnetic-ferromagnetic transition continues to be of second order as in the ordinary Ising model. As Δ is increased one sees a line of critical points (the solid line in Fig. 1.7), until beyond a critical value Δ_c, the transition changes into a first order one and the cross-over occurs at what is called a tricritical point [35].

As Δ is increased still further, the transition temperature decreases and goes to zero. This is so because, for a large value of Δ the $S_i = \pm 1$ levels become inaccessible and hence the exchange interaction becomes inoperative.

The following features about the various terms in the Hamiltonian in Eq. (1.15) are worth noting in our understanding of tri-critical behaviour. First, one has the usual magnetic order $M = <S_i>$ which characterizes the ferromagnetic to paramagnetic transition. In addition, one has, for this spin-1 system, a quadrupolar order parameter $Q = <S_i^2>$. Note that the field Δ which is conjugate to Q plays the same role as H does with respect to M. Further, the order parameters M and Q are not totally

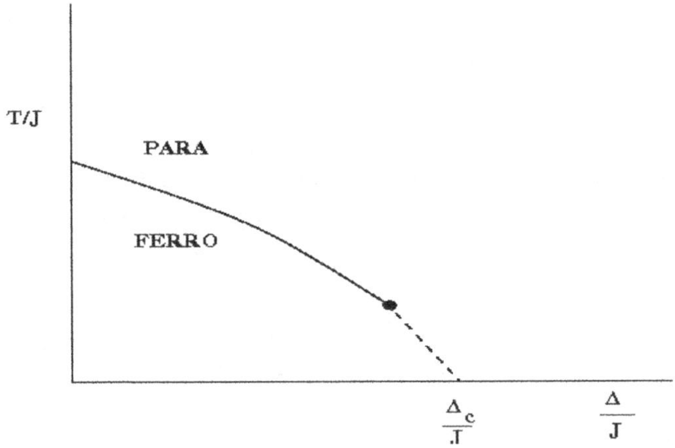

Fig. 1.7. Phase diagram in the $T - \Delta$ plane. For $\Delta < \Delta_c$, one has a second order transition from the para-to-ferro-magnetic phase. For $\Delta > \Delta_c$, the transition changes over into first order (dashed line).

independent of one another. For example, it is not possible to have a stable state in which $M = 1$ and $Q = 0$, since $M = 1$ requires that all the spins have $S_i = 1$ which in turns leads to $Q = 1$. The two-order parameters are therefore kinematically coupled in the sense that even in the absence of interactions, there are constraints on the values which can be taken on by the two together. It is the competition between the two kinematically coupled order parameters which is eventually responsible for the occurrence of the tricritical point. The exchange term in Eq. (1.15) gives a lower energy for $M = 1$, while the crystal field term favours $Q = 0$. These two values are incompatible, however. The increase in Δ ultimately makes it more favourable energetically to have $Q = 0$ (and hence $M = 0$) than $M = 1$.

The meaning of the various phases, that are expected to occur in relation to Fig. 1.7, become clearer if one considers an applied magnetic field H in addition to the crystal field and Ising terms in Eq. (1.15). The Hamiltonian now becomes

$$\mathcal{H} = -J\sum_{<ij>} S_{iz}S_{jz} + \Delta\sum_{i} S_{iz}^2 - H\sum_{i} S_{iz}\,. \tag{1.16}$$

The corresponding phase diagram, in the space of three field variables (in the sense of Sec. 1.2): T, Δ and H, is shown in Fig. 1.8. The two-dimensional phase diagram shown in Fig. 1.7 is a projection of the three-dimensional diagram of Fig. 1.8 onto the $H = 0$, $(T - \Delta)$ plane. One now has a first order co-existence surface A in the $H = 0$ plane, on which the order parameter M assumes nonzero values with opposite signs depending on whether $H \to 0$ from positive or negative directions, exactly as in Fig. 1.7, for $T < T_c$. As T increases, keeping Δ 'small' the surface A terminates in a line of critical points (the bold solid line on the $H = 0$ plane in Fig. 1.7). However, for $T < T^*$ and $\Delta > \Delta^*$, this line of critical points terminates in the two-phase co-existence line D, which is a line of first order transitions, shown as a dashed-dotted line in Fig. 1.8. Looking at Fig. 1.8 one notes that A is connected along the line D to two first order (co-existence) surfaces B and B' extending symmetrically into the regions $H > 0$ and $H < 0$, respectively. The surfaces B and B' themselves terminate with increasing temperature in lines (bold solid ones) of critical points. These two critical lines (for $H \neq 0$) fan out symmetrically from the $H = 0$ plane and join the critical line (for $H = 0$) at the tricritical point (the black dot). Thus, the tricritical point may be regarded either as the termination of a line (D) of triple (or three-phase co-existence) points, or equivalently, as the confluence of three lines of critical point. In fact, it is the latter feature which prompted Griffiths to name the special point a tricritical point [35]. We will return to further discussion of the phase diagram in Fig. 1.8 in the context of a very different system, viz. a $He^3 - He^4$ mixture.

A model of the kind described by Eq. (1.15) has been used by Blume to explain first-order magnetic phase changes in UO_2 [33]. It was found that depending on the ratio of the crystal field splitting Δ to the molecular-field splitting (appropriate to the exchange term in Eq. (1.15)), one obtains either a paramagnetic state, a first order phase change (indicated by the dashed line in Fig. 1.7), or a second order transition (shown as the solid line in Fig. 1.7). The confluence of the first order line and the second order line, viz. the black dot in Fig. 1.7, is a tricritical point, as discussed above. While the tricritical phenomenon in UO_2 takes place in the absence of an external magnetic field, first-order phase transitions are known to occur in magnets to which an external magnetic field is applied. Such systems are known as meta-magnets. A case in point is the system of $FeCl_2$ which belongs to a class of meta-magnets that exhibit a tricritical point [36]. The mechanism is discussed below.

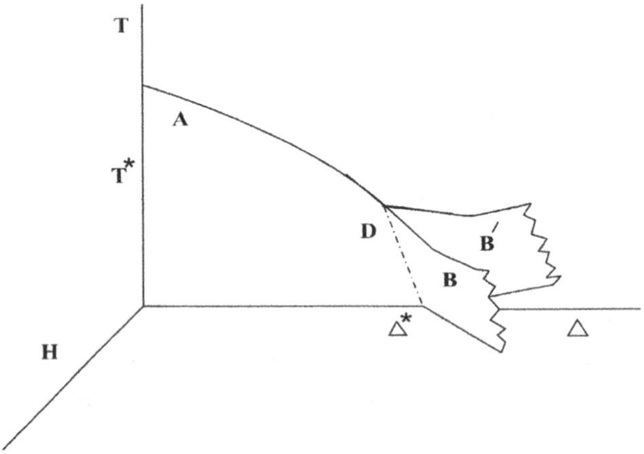

Fig. 1.8. Phase diagram for an Ising magnet in the presence of a crystal field Δ and a magnetic field H. There are three surfaces of first order transitions A, B, and B', which meet at the tricritical point that may be viewed also as the meeting point of three lines (solid) of second order phase transitions.

In $FeCl_2$, Fe ions are placed on stacks of parallel triangular lattices separated by Cl ions, as shown in Fig. 1.9. Each Fe ion is surrounded by a cage of six Cl ions. The iron is in the valence state of Fe^{2+} and carries an effective spin of one. The sign of the exchange interaction within a layer is positive, favouring ferromagnetism whereas that between layers is negative yielding anti-ferromagnetism. Further, the spins are preferentially aligned perpendicular to the layers because of lattice anisotropy. Thus, in the fully ordered state, the spins in a given plane are parallel, but they alternate in direction from layer to layer, thereby yielding a two-sub-lattice anti-ferro-magnetic order. When an external magnetic field is applied normal to the layers, the sub-lattice magnetization which is conducive to the applied field will show enhancement, whereas the neighbouring sub-lattice magnetization will undergo a decrease. Thus, there occur both a uniform magnetization and

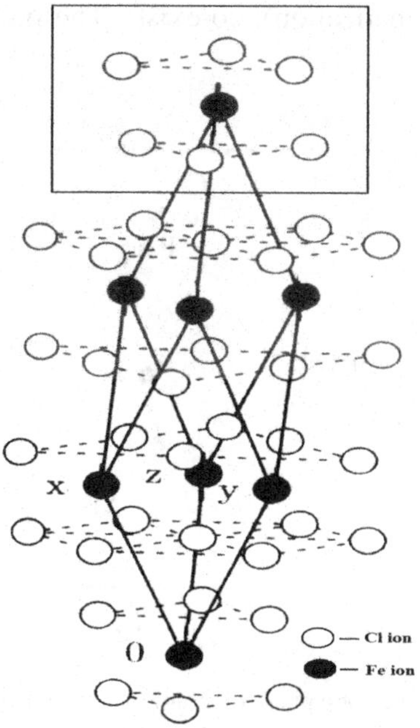

Fig. 1.9. Crystal structure of $FeCL_2$.

a sub-lattice magnetization. Again, we have a situation with two competing order parameters, like in the example discussed in the beginning of this section. When the applied field is large enough to polarize all the spins, anti-ferro-magnetic order will be destroyed.

The experimental phase diagram for $FeCl_2$, for which the uniform magnetization m is plotted along the ordinate and the temperature T is plotted along the abscissa, is shown in Fig. 1.10. When T is large and the uniform magnetization is sufficiently small we are in the anti-ferro-magnetic (AF) phase. Keeping T fixed on the high temperature side and enhancing m, as can be achieved by increasing the strength of the applied magnetic field, one crosses into the paramagnetic (PARA) phase, as argued above. The line of demarcation is a line of critical points i.e. a line of second order phase transition. Below a certain temperature indicated by a black dot and labeled by T_{tr}, we have phase separation, meaning thereby that two phases, one with 'large' uniform magnetization and the other with small uniform magnetization (and concomitantly, a 'large' staggered magnetization), co-exist. The point T_{tr} is the tricritical point.

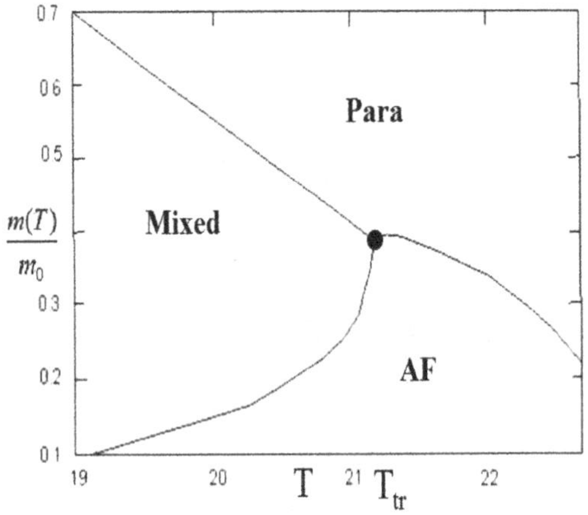

Fig. 1.10. Schematic phase diagram for the metamagnet $FeCl_2$.

It may be noted that the phase diagram in Fig. 1.10 is in the mixed density-field variables space, in accordance with the nomenclature introduced in Sec. 1.2. Thus the portion of the diagram to the left of T^* looks qualitatively like the magnetization versus temperature diagram of an ordinary ferro-magnet. Had we drawn the phase diagram in the field variables space by replacing the magnetization by its conjugate magnetic field the two-phase co-existence region would have shrunk to a first order line, meeting the second-order line at the tricritical point, as in Fig. 1.7.

A very similar phenomenon of tricritical points is found in a completely different system of a binary mixture of He^3 and He^4 in which He^4 can occur in two different phases: normal and suverfluid, the two phases being related by a symmetry operation. Indeed a slight modification of the Hamiltonian as employed in Eq. (1.16) for the meta-magnet $FeCl_2$ goes under the name of the Blume-Emergy-Griffiths (BEG) model that was used for describing the tricritical point in He^3 – He^4 mixtures [37]:

$$\mathcal{H} = -J\sum_{<ij>} S_{iz}S_{jz} - K\sum_{<ij>}(S_{iz})^2(S_{jz})^2 + \Delta\sum_i S_{iz}{}^2 - H\sum_i S_{iz}. \quad (1.17)$$

Under saturated vapour pressure it is observed that the superfluid λ-transition temperature decreases with increasing mole fraction x of He^3, and below a temperature $T^* = 0.87K$ corresponding to $x = 0.67$, a first order phase separation takes place (Fig. 1.11).

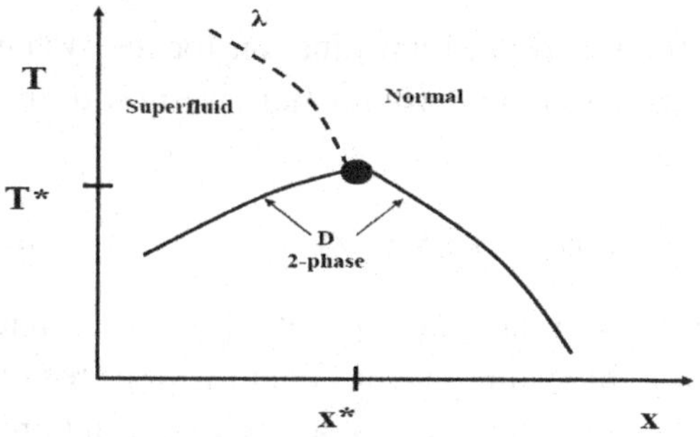

Fig. 1.11. Phase diagram (schematic) for $He^3 - He^4$ mixtures. The two-fluid coexistence curve is labeled D, the dashed curve is the line of the lambda transitions and the dot indicates the tricritical point.

The corresponding phase diagram in the space of field variables is the same as in the case of the metamagnet $FeCl_2$ (Fig. 1.8) in which Δ may be interpreted as the themodynamic field conjugate to x, i.e., Δ is the difference in chemical potentials of He^3 and He^4, and H is a fictitious field conjugate to the superfluid order parameter ψ. The first order coexistence surface A in the $H = 0$ plane extends to $\Delta = -\infty$ (pure He^4). On A, ψ assumes nonzero values with opposite signs, as described earlier following Eq. (1.16). Thus ψ is the analog of the magnetization M, the order parameter in the magnetic case.

The BEG model has a built-in time-reversal symmetry (in the absence of H) with the consequence that odd terms in the spin operator S_{iz}, for instance, are absent. The most general spin-one model however, which subsumes the BEG model, Eq. (1.17), and the Hamiltonian for the metamagnet, Eq. (1.16), does not have to obey this symmetry, and is therefore, applicable to a ternary liquid mixture. The corresponding Hamiltonian can be written as [38]:

$$\mathcal{H} = -J\sum_{<ij>} S_{iz}\,S_{jz} - K\sum_{<ij>} (S_{iz})^2\,(S_{jz})^2 - C\sum_{<ij>} [(S_{iz})^2\,S_{jz} + S_{iz}(S_{jz})^2] + \Delta\sum S_{iz}{}^2 - H\sum S_{iz}. \qquad (1.18)$$

Solving for the grand partition function for the system described by Eq. (1.18) in the mean field approximation, we may derive for the Gibbs potential:

$$\mathrm{G} = \bar{\mathrm{a}}\mathrm{y}\mathbf{z} + \bar{\mathrm{b}}\mathrm{x}\mathbf{z} + \bar{\mathrm{c}}\mathrm{xy} + \ln\mathrm{x} + \ln\mathrm{y} + \ln\mathbf{z}\,, \qquad (1.19)$$

where the variables x, y and $\mathbf{z}(= 1 - x - y)$ can be interpreted as the mole fractions of a three component system whose free energy, as given by Eq. (1.19), has the structure of a "regular solution" model. The "energy" parameters $\bar{a}$, $\bar{b}$, and $\bar{c}$, are given in terms of the Boltzmann constant k_B, the temperature T and the parameters appearing in the Hamiltonian in Eq. (1.18):

$$\bar{a} = \frac{1}{2\mathrm{k}_B T}(J + K - 2C), \bar{b} = \frac{1}{2\mathrm{k}_B T}(J + K + 2C), \bar{c} = \frac{2J}{\mathrm{k}_B T} \qquad (1.20)$$

For $\bar{a} = \bar{b} = \bar{c}$, G is invariant under the permutation of x, y and $\mathbf{z}$. Consequently one finds, in addition to tricritical points, lines of tricritical points, four-phase coexistence region and three fourth order points at which four coexisting phases become simultaneously identical.

1.4. Multicriticality in an Open System of Two-Mode Lasers

All the critical and multicritical phenomena that we have discussed so far pertain to systems in thermal equilibrium, the properties of which are described by the rules of ordinary Thermodynamics. Nature however abounds with a variety of systems in different disciplines of biology, chemistry and physics which exhibit behavior akin to phase transitions from disordered to ordered states. These systems, in contrast to the thermodynamic ones, are 'open' systems in which the usual control parameters e.g., temperature or pressure have no role to play, and are maintained in steady states under the influence of certain 'pump' fields. It is the analogy between the properties of these steady states with thermodynamic equilibrium states that has led to the study of critical and multicritical phenomena in these open systems. Examples are coherent lasing properties of a single-mode laser in the presence of an electromagnetic pump, oscillating chemical reactions under the infusion of a reactant, population dynamics in predator-prey models, etc. [39].

Given this background we want to extend the analogy between phase transitions in magnets and lasers, and discuss in this section the occurrence of multicritical points in two-mode lasers, such as the Zeeman or the ring laser [40]. The basic equations are in the form of Langevin equations for the complex amplitudes $\in_1$ and $\in_2$ associated with the two modes of oscillation:

$$\dot{\in}_1 = (a_1 - |\in_1|^2 - \xi |\in_2|^2)\in_1 + f_1(t), \qquad (1.21)$$

$$\dot{\in}_2 = (a_2 - |\in_2|^2 - \xi |\in_1|^2)\in_2 + f_2(t), \qquad (1.22)$$

where a_1 and a_2 are dimensionless pump parameters that could be either positive or negative and ξ is a (positive) coupling parameter between the two modes. The complex quantities $f_i(t)$ are Gaussian, delta-correlated random noise terms with zero-mean, having their origin in electromagnetic field fluctuations, mirror-loss, etc.:

$$< f_i^*(t) f_j(t') \geq 2\,\delta_{ij}\;\delta(t - t'). \qquad (1.23)$$

While the above set of equations leads to fully time-dependent solutions for the random variables $\epsilon_i(t)$, driven by the stochastic fields $f_i(t)$, the steady-state time-independent properties can be ascertained from an underlying probability distribution that can be deduced as

$$P_{st}(\epsilon_1, \epsilon_2) \propto \exp(-\Phi), \qquad (1.24)$$

where Φ has the formal structure of a Landau free energy:

$$\Phi = -\frac{1}{2}a_1|\epsilon_1|^2 - \frac{1}{2}a_2|\epsilon_2|^2 + \frac{1}{4}(|\epsilon_1|^4 + |\epsilon_2|^4) + \frac{1}{2}\xi|\epsilon_1|^2|\epsilon_2|^2. \tag{1.25}$$

It is this free energy that is formally analogous to that of an anisotropic antiferromagnet that displays bicritical and tetracritical points.

The phase diagrams are obtained from

$$\frac{\delta\Phi}{\delta\epsilon_1} = \frac{\delta\Phi}{\delta\epsilon_2} = 0, \tag{1.26}$$

and the 'local' stability of the phases is assessed from the positive-definitenes of the determinant of the Hessian matrix:

$$D = \begin{vmatrix} \dfrac{\delta^2\Phi}{\delta\epsilon_1\,\delta\epsilon_1^*} & \dfrac{\delta^2\Phi}{\delta\epsilon_1\,\delta\epsilon_2^*} \\ \dfrac{\delta^2\Phi}{\delta\epsilon_2\,\delta\epsilon_1^*} & \dfrac{\delta^2\Phi}{\delta\epsilon_2\,\delta\epsilon_2^*} \end{vmatrix} \tag{1.27}$$

The phase diagram, for $\xi < 1$, exhibiting a tetracritical point, is shown in Fig.1.12. On the other hand, when the coupling constant ξ switches to a value larger than one ($\xi > 1$) the phase characteristics change (Fig. 1.13). One now has a bicritical point B at the confluence of two lines of second order transition (heavy lines) and a line of first order transition (dashed line).

The occurrence of a bicritical point is the consequence of the competition between two order parameters ($I_1 = |\epsilon_1|^2$ and $I_2 = |\epsilon_2|^2$, in the present instance) in complete analogy with the phenomenon observed in a weekly anisotropic (uniaxial or orthorhombic) anti-ferromagnet in a magnetic field applied along the easy direction of antiferromagnetic order. As shown by Fisher and Nelson (cited in [40])

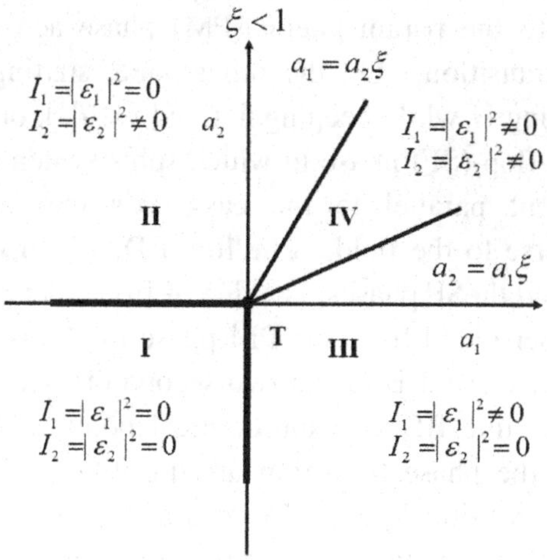

Fig. 1.12. Phase diagram displaying four possible phases (I – IV) in the space of the pump parameters a_1 and a_2. For the coupling constant $\xi < 1$, four (heavy) lines of 2nd order phase transition meet at T called the tetracritical point.

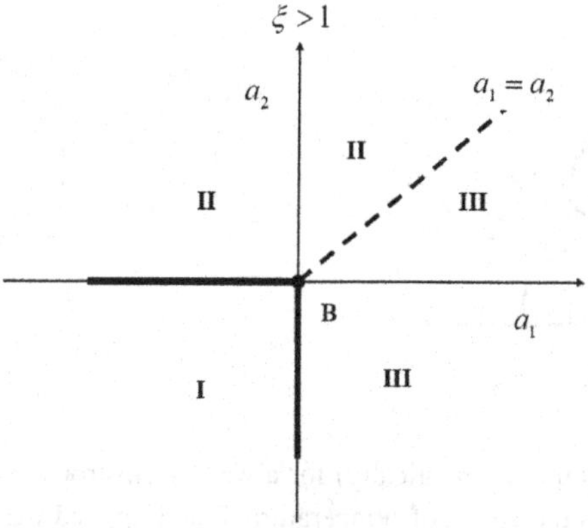

Fig. 1.13. Two (heavy) lines of second order transition meet one (dashed) line of first order transition at the point B, termed the bicritical point.

the system is in the antiferromagnetic (AF) phase characterized by "up" – "down" ordering of the spins, below a certain temperature T and the magnetic field H (Fig. 1.14). If one keeps H fixed but increases T one goes over into the paramagnetic (PM) phase across the line AB of second-order transition. On the other hand, starting from the AF phase and increasing H while keeping T fixed ($< T_B$), one marches into the so-called spin-flop (SF) phase in which spins switch over from their low-field alignment parallel to the easy axis into a perpendicular alignment transverse to the field. The line BD, which is the boundary between the AF and theSF phases, is a line of first order phase transition. The SF phase is separated from the PM phase by the second-order line BC. The confluence point B of the two second-order lines AB and BC and the first order line BD is a multicritical point at which the basic characteristics of the phase transition are quantitatively different from that near an ordinary critical point. Comparing Fig. 1.13 and the inset of Fig. 1.14 we find that a qualitatively similar situation exists in the case of a two-mode laser for which Phase I, located symmetrically with respect to phases II and III, plays the role of the paramagnetic phase.

Concluding, we must add that the case $\xi < 1$, yielding a tetracritical point, also has its counterparts in antiferromagnetism.

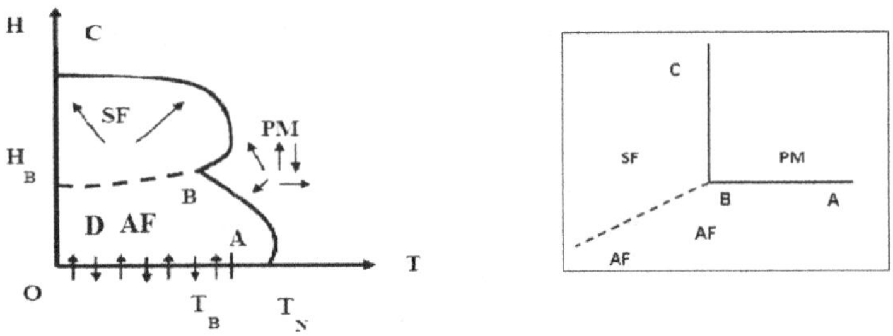

Fig. 1.14. Phase diagram (schematically) for a weakly anistropic antiferromagnet in the thermodynamic space of temperature T and applied magnetic field H. The Neel point T_N marks the antiferromagnetic (AF) – paramagnetic (PM) transition for H =0. The point B(H_B, T_B) at which the AF, PM and the spin-flop (SF) phases meet is referred to as the bicritical point. The blown-up inset on the right shows the three lines of phase transitions in the region asymptotically close to the bicritical point B.

Chapter 2

Quantum Phase Transition: Transverse Ising Model and Other Systems

2.1. Introductory Remarks

The concept of the order parameter as a thermodynamic quantity to characterize a phase transition, defined by the statistical average of a dynamical variable, was already introduced in Chapter I. In an Ising model for instance, the appropriate dynamical variable is $< S(\vec{R}) >$ while its average $< S(\vec{R}) >$ is the order parameter m $(\vec{R})$ measured at the position $\vec{R}$. An important attribute of the order parameter is its fluctuation property measured by the correlation function $c(\vec{r}) = < S(\vec{R})S(\vec{R}+\vec{r}) >$ and the associated correlation length ξ, defined by* $C(\vec{R}) = exp\,(-(\vec{r})/\xi$. Just as ξ is a measure of spatial fluctuations, a related quantity is the correlation time τ that quantifies temporal correlations, characterized by the auto (i.e. same point) – correlation function: $C(\tau) =< S(\vec{R},t')S(\vec{R},t'+t) >= exp(-t/\tau)$. One of the fascinating features of phase transitions is that both ξ and τ diverge at the critical point, for instance the Curie temperature, which demarcates the paramagnetic and the ferromagnetic phases of a magnet. This divergence is associated with the fact that fluctuations of the order parameter become infinite-ranged at the critical point, in the spatial as well as temporal senses.

Against this background it is important to assess the relative strength of the correlation time τ, or more logically, its inverse called the 'correlation frequency' ω, vis-à-vis an inherent frequency scale ω_0

*This form of the correlation function is valid away from a critical point of phase transition.

governed by the thermal energy $k_B T$ divided by the Planck constant $\hbar(\omega_o \equiv k_B T/\hbar)$. As the correlation time τ becomes infinity at the critical temperature T_c, the correlation frequency ω vanishes whereas ω_0 remains finite as long as T_c stays finite. Therefore, $\omega_c(= k_B T_c/\hbar) > \omega_0\ (T = T_c)$, for all finite – temperature phase transitions. What then is the significance of the frequency scale ω_0? Recall that Quantum Mechanics is the operative mechanism that governs the dynamics of a system at zero temperature. At that temperature our system of interest cannot exchange any energy with its environment and therefore the quantum wave function of the system maintains a definite relation at two different times, because the underlying evolution in time is unitary. This implies that the phase of the wave function remains coherent, a property that is responsible for such remarkable phenomena as interference. When the temperature is lifted from zero the system begins to undergo energy–exchanges with its environment through various interaction terms. As a result effect $(k_B T)/(=\omega_0^{-1})$s of quantum coherence start getting weaker until at a sufficiently high temperature the system become completely incoherent. How does one characterize this coherence to decoherence transition? Well, the quantity, $\hbar/k_B T/(= \omega_0^{-1})$ sets a time-window within which quantum coherent properties are maintained and naturally, this time–window is infinitely wide at precisely zero temperature. As the temperature becomes larger the time window becomes progressively narrower, and ultimately at a sufficiently high temperature determined by system properties, there is no time scale left within which the coherence effects can be detected. Appropriately therefore, ω_0 may be referred to as the 'quantal frequency'.

Now, for a system undergoing phase transition at a finite temperature the quantal frequency at T_c remains finite and larger than the corresponding critical frequency ω_0 . Hence, on time scales over which the statistical fluctuations of the underlying order parameter are relevant, the system has already lost its quantum coherence properties. This implies that for such systems phase transition properties can be treated by classical methods – a remarkable observation that indeed is the underlying reason why all finite temperature phase transitions can be handled by essentially classical theories such as Landau – Ginzburg phenomenology [41]. The above-statement is true even for systems such as a super-conductor or a superfluid though the very existence of the order parameter in these systems is rooted in quantum mechanics, if one notes that the condensate wave function is, after all a coherent 'extended state'.

All the above considerations are of course invalid if the transition temperature T_c is zero. Then, at the critical point, the coherence time window is infinitely large and hence classical methods fail in providing a proper theoretical description of what are understandably called the quantum phase transitions. However, the question arises: if the temperature is frozen to zero and is not available as a disordering field to tune the phase transition, what is the appropriate control parameter? The answer to this question and indeed to the issue of quantum phase transition and the concomitant quantum critical phenomenon can be most simply (and clearly) found in a magnetic paradigm by slightly generalizing the Ising model.

Consider then Eq. (1.14) with an additional control parameter Γ that couples to the x-component of the total spin leading to the Hamiltorian:

$$\mathcal{H} = -\sum_{<ij>} J_{ij} S_{iz} S_{jz} + \Gamma \sum_i S_{ix} \; . \tag{2.1}$$

Quite naturally, the model described by Eq. (2.1) is referred to as the Transverse Ising Model (TIM) in which Γ is proportional to a magnetic field applied in a direction transverse to the easy Ising Axis (Z, in this instance) [42]. While the order parameter for the ordinary Ising Model ($\Gamma = 0$) has thermodynamic properties only, that for the TIM has both thermodynamic and dynamic properties inexorably linked-up, as the operator S_{iz} no longer commutes with the Hamiltonian $\mathcal{H}$. This has the following important consequences. Consider the system at zero temperature. When the transverse field Γ is absent the spins are all aligned along the Z-axis yielding a fully ordered ferromagnetic state. As Γ is switched on, there is an additional torque on the Z-component of the spin, the quantization axis is tilted away from the Z-axis, and the order parameter decreases. In the quantum mechanical sense, the transverse term, which couples to S_{ix}, causes transitions between the eigenstates of the operator S_{iz}. Eventually, when Γ exceeds a certain critical value $\Gamma_{c,}$ the quantization axis loses its meaning, the spin-alignment is completely disrupted and as a consequence, the order parameter vanishes. Thus, in the absence of temperature, Γ is now the new control parameter – a disordering field that can be tuned to effect the above ferromagnetic to paramagnetic transition. But, for reasons mentioned above, this zero-temperature phase transition can no longer be described by classical theories – we have a quantum phase transition (QPT) in which Γ_c marks a quantum critical point (QCP). While a mean-field theory for the QPT will be presented in Sec. 2.2 suffice it to note that the scaling behaviour near the QCP belongs to a universality class that is different from that near the ordinary critical point T_c.

In Sec. 2.3 below we shall introduce a rare earth magnet $LiHoF_4$ which is a perfect realization of QPT. In addition, the mean field theory described in Sec. 2.2 works so extremely well in $LiHoF_4$ that the latter has been dubbed as the model quantum magnet. Thus, Magnetism once again provides a neat platform on which the disordering field can be easily realized in the laboratory, the quantum critical properties can be studied in a controlled manner and the general features of QPT can be analyzed rigorously, which have a bearing on numerous other phenomena [43]. Consider then the Metal-Insulator transition (at zero temperature) in which a metal loses its electron conduction property as a result of defects that lead to electron-localization. The appropriate field variable is thus a chemical potential which is thermodynamically conjugate to the defect concentration but unlike Γ for the TIM this field is not physically realizable in the laboratory, although a whole lot about the metal-insulator transition can be learnt by studying the TIM. Other examples of QPT that we will discuss a bit in Sec. 2.4 can be found in arrays of Josephson junctions [44].

2.2. The Mean Field Theory of TIM

In the mean field approximation the many body problem is reduced to a tractable one body problem by assuming the spin at a given site i to be embedded in an effective medium that creates a local magnetic field H_i at the site i. The Hamiltonian in Eq. (2.1) can then be written as

$$\mathcal{H}_i = -(H_i S_{zi} + \Gamma S_{xi}), \tag{2.2}$$

where the field H_i is approximated as

$$H_i = \sum_{j=1}^{z} J_{ij} < S_{zj} >, \tag{2.3}$$

z being the number of nearest neighbour sites. Furthermore, if we assume the system to be uniform, $< S_{zj} >$ is independent of the site index j. Therefore $H_i \equiv H = J(0) < S_z >$, where

$$J(0) = \underset{j=1}{\overset{z}{Z}} J_{ij}. \tag{2.4}$$

Hence, Eq. (2.2) yields

$$\mathcal{H}_i = -(HS_{zi} + \Gamma S_{xi}). \tag{2.5}$$

The partition function is given by

$$Z_i = Tr(e^{-p\mathcal{H}_i}) = Tr[e^{\beta(HS_z + (\Gamma S_x)}], \tag{2.6}$$

where $\beta = (K_B T)^{-1}$. Because S_z takes only two discrete values +1 and −1 we can use properties of 2x2 matrices to decompose the exponential operator in Eq. (2.6), in order to simplify Z_i as

$$Z_i = 2\cos\hbar(\beta h), \tag{2.7}$$

independent of i and where

$$h = \sqrt{H^2 + \Gamma^2}. \tag{2.8}$$

Thus

$$< S_z >= \frac{1}{2\cosh(\beta h)} Tr[S_z^{\beta(HJ_z + (\Gamma S_x)}], \tag{2.9}$$

which yields

$$< s_z >= \frac{H}{h}\tanh(\beta h). \tag{2.10}$$

Similarly,

$$< S_x >= \frac{\Gamma}{h} \tanh(\beta h). \tag{2.11}$$

Denoting the LHS of Eqs. (2.10) and (2.11) as the components of the magnetization m_z and m_x respectively, we obtain the transcendental equations:

$$m_z = \frac{J(o)m_z}{\sqrt{(J(o)m_z)^2 + \Gamma^2}} \tanh(\beta\sqrt{(J(o)m_z)^2 + \Gamma^2}), \tag{2.12}$$

$$m_x = \frac{\Gamma}{(J(o)m_z)^2 + \Gamma^2} \tanh \beta(\beta\sqrt{(J(o)m_z)^2 + \Gamma^2}). \tag{2.13}$$

The above equations yield self-consistent relations for the two relevant magnetization components. Note that the order parameter m_z vanishes when either Γ or T goes to infinity. This confirms our earlier physical argument that Γ acts as a disordering field.

We may now discuss the phase diagram predicted by Eqs. (2.12) and (2.13) by designating the order parameter m_z as ψ and defining the critical temperature for the pure $(i.e.\ \Gamma = 0)$ Ising model as

$$T_c^o = \frac{J(o)}{k_B} \tag{2.14}$$

Equation (2.12) then yields (for the solution corresponding to $\psi \neq o$)

$$1 = \frac{T_c^o / T}{\sqrt{(T_c^o \psi / T)^2 + \bar{\Gamma}^2}} \tanh\left(\sqrt{\left(\frac{T_c^o}{T}\psi\right)^2 + \bar{\Gamma}^2}\right), \tag{2.15}$$

where $\bar{\Gamma} = \beta\Gamma$.

At the phase boundary, $\psi = o$, thus leading to

$$T_c = \frac{T_c^o}{\bar{\Gamma}_c} \tanh \bar{\Gamma}_c, \tag{2.16}$$

which is equivalent to

$$\frac{\Gamma}{J(o)} = \tanh\left(\frac{\Gamma}{K_B T_c}\right). \tag{2.17}$$

The corresponding phase diagram is shown in Fig. 2.1 demarcating the following three regimes:

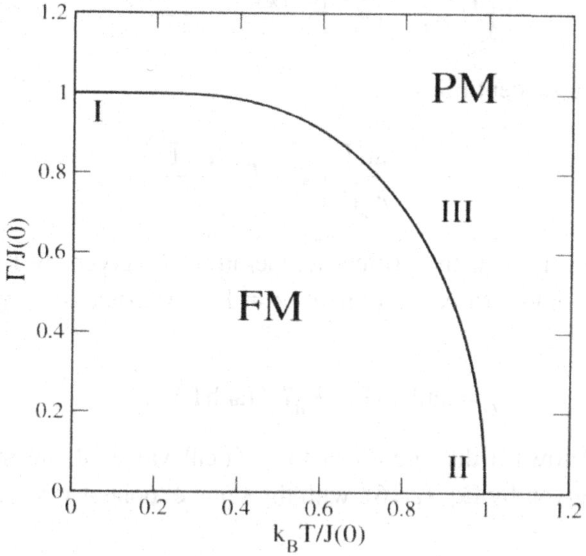

Fig. 2.1

<u>Region I</u>, where Γ is <u>large</u>, and T is <u>small</u> – quantum effects are dominant here.
<u>Region II</u>, where Γ is <u>small</u> – classical effects are dominant and the TIM is analogous to the usual Ising model.
<u>Region III</u>, where both Γ and T are moderate.

We next analyze the static susceptibility in region III, by adding a small aligning field ΔH along the Z-axis. In that case, Eq. (2.15) can be rewritten as

$$\psi = \frac{\beta\Delta H + T_c^o \psi/T}{\sqrt{(\beta\Delta H + T_c^o \psi/T)^2 + \bar{\Gamma}^2}} \tanh\left[\sqrt{(\beta\Delta H + T_c^o \psi/T)^2 + \bar{\Gamma}^2}\right]. \tag{2.18}$$

Linearzing in ψ and ΔH leads to

$$\psi\left(1 - \frac{T_c^0}{T}\frac{\tanh}{\bar{\Gamma}^2}\right) = \beta\Delta H \frac{\tan \hbar\bar{\Gamma}}{\bar{\Gamma}}, \tag{2.19}$$

yielding for the susceptibility

$$\chi = \frac{\psi}{\Delta H} = \frac{\tan\bar{\Gamma}}{k_B\bar{\Gamma}}\left(T - T_c^0 \frac{\tan\bar{\Gamma}}{\bar{\Gamma}}\right)^{-1}. \tag{2.20}$$

Thus, χ diverges at the critical temperature T_c given by Eq. (2.16) with a mean field critical exponent $\gamma = 1$. Alternatively, χ can be rewritten as

$$\chi = \tanh\bar{\Gamma}(\Gamma - K_B T_c^o \tanh\bar{\Gamma})^{-1}, \tag{2.21}$$

which again shows a divergence at the critical value of the transverse field $\Gamma = \Gamma_c$ given by Eq. (2.16) with the same exponent $\gamma = 1$.

2.3. The Quantum Magnet LiHoF$_4$

The system $LiHoF_4$ belongs to a class of compounds $LiRF_4$, where R is a rare earth element (Fig. 2.2). Although the rare earth elements of Ho^{3+} in $LiHoF_4$ are coupled by dipole-dipole interaction, only the lowest doublet of the 17 crystal-field-split states is appreciably populated at low temperatures ($T < 2K$) because of a large crystal field. Further, the sign of the crystal field is negative; hence the lowest doublet corresponds to the highest spin states. Consequently, the off-diagonal terms of the dipolar interaction are effectively "quenched" leading to the so-called "truncated dipolar Hamiltonian" [45]. The latter is of the Ising

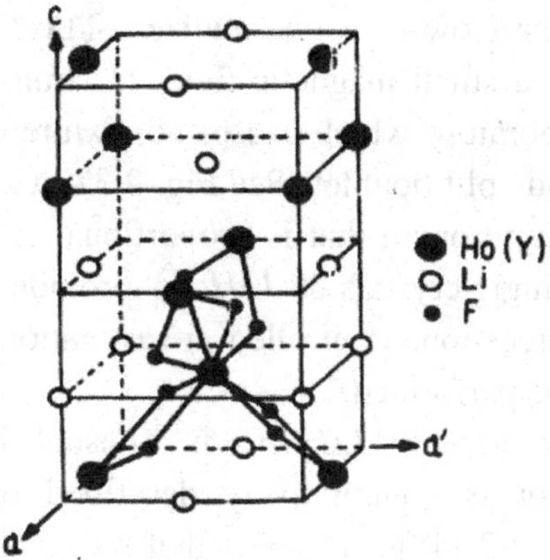

Fig. 2.2. The Crystal structure of $LiHoF_4$.

type, yielding an Ising coupling between spins projected along the crystallographic c-axis, with the interaction J_{ij} of the form:

$$J_{ij} = J\left(\frac{1-3\cos^2\theta ij}{r_{ij}^3}\right), \tag{2.22}$$

where θ_{ij} is the angle between the Ising axis (i.e. the c-axis of the crystal) and the position vector connecting the spins. As seen in Eq. (2.22), J_{ij} is very long-ranged because of its r_{ij}^{-3} dependence on the inter-spin distance r_{ij}. This indicates that a mean-field treatment will be in order for such a system. However, the angular dependence $[1-3\cos^2(\theta_{ij})]$ allows for J_{ij} to change sign, implying that the interaction switches from ferromagnetic ($J_{ij} > o$) to antiferromagnetic ($J_{ij} < 0$) for intermediate angles ($55^\circ \le \theta_{ij} \le 125^\circ$). While this occurrence of competing interactions may raise doubts about the uniqueness of the ground state, a classic analysis by Luttinger and Tisza [46] demonstrates that the ground state is indeed that of a dipolar ferromagnet for the b.c.c structure of $LiHoF_4$. The experimentally measured critical temperature for the paramagnetic – ferromagnetic transition turns out to be $T_c = 1.53K$.

Having arrived at a situation in which one has an Ising model as an approximate low-temperature description of the dipolar-coupled system, the question arises: How does one realize a TIM? In the experiment of Bitko et al. [47], a small magnetic field H_t transverse to the c-axis is applied in the laboratory which causes admixture of the two eigenstates of the crystal-field-split doublet (See Fig. 2.3), yielding (in perturbation theory) a spin Hamiltonian that is proportional to S_x, with a prefactor $\Gamma \infty H_t^2$. Thus, single crystals of $LiHoF_4$ provide a clean model system in which high-precision, controlled investigation of quantum critical phenomena can be performed.

In one other aspect, $LiHoF_4$ is unusual in that the observed magnetic behavior is quantitatively described by mean-field theory, discussed in Sec. 2.2 [48]. The fact that the quantum critical behavior in d = 3 is mean field-like is not surprising as it is known that the d-dimensional quantum phase transition (at $T=0$) for the TIM is equivalent to the $(d+1)$ dimensional thermal phase transition for the ordinary Ising model [49, 50]. The mean field nature of the thermal phase transition observed in $LiHoF_4$ is also not surprising as renormalization group (RG) theory shows that, for a dipolar-coupled Ising system, the marginal dimensionality is $d_C = 3$, barring logarithmic corrections for critical exponents. [51]. (Recall that we expect mean field behavior for $d > d_c$.)

We now bring-in one additional attribute of the $LiHoF_4$ system that is crucial for a satisfactory understanding of the experimentally observed phase diagram, especially at low temperatures. The ionic spin of Ho^{3+} element has a non-negligible hyperfine interaction with the nuclear spin of Ho in which the coupling constant is of the order of 0.039K. Because the large crystal field 'quenches' the off-diagonal terms of the spin operator of Ho^{3+} (with the crystallographic c-axis as the direction of quantization), as mentioned before, the hyperfine interaction is also axial [48]. Incorporation of this interaction generalizes the Hamiltonian in Eq. (2.1) to:

$$\mathcal{H} = -\sum_{<ij>} J_{ij} S_{iz} S_{jz} - \Gamma \sum_{i=1}^{N} S_{ix} - a \sum_{i=1}^{N} I_{iz} S_{iz}, \tag{2.23}$$

where 'a' is the hyperfine coupling constant and I_{iz} is the nuclear spin of the ith Holmium nucleus. Correspondingly the single site mean field Hamiltonian in Eq. (2.2) has to be generalized to:

$$\mathcal{H}_i = -[(H_i + aI_{zi})S_{zi} + \Gamma S_{xi}]. \tag{2.24}$$

With the Hamiltonian in Eq. (2.24) we now proceed to evaluate the phase diagram [52]. In accordance with our earlier analysis the partition function for a single site in the mean field approximation can be written as

$$Z = \left[Tre^{\beta((H+aI_z)S_z+\Gamma S_x)}\right], \tag{2.25}$$

independent of i and in which the trace is over the ionic as well as the nuclear spin eigenstates of *Ho*. Labeling the 8 nuclear eigenstates by $|M>$, we have $I_z|M>= M|M>$ where M assumes values – 7/2, -5/2, ...5/2, 7/2.

Thus

$$Z = \sum_{M=-7/2}^{+7/2} 2\cosh \beta h(M), \tag{2.26}$$

where

$$h(M) = \sqrt{(H + aM)^2 + \Gamma^2}. \tag{2.27}$$

Proceeding as before, Eq. (2.12) for the order parameter can be written as

$$m_z = \frac{\sum_M [\frac{aM + J(o)m_z}{h(M)}]Sinh(\beta h(M))}{\sum_M \cosh(\beta h(M))}. \tag{2.28}$$

The phase boundary shown in Fig. 2.3 is estimated from a numerical solution of Eq. (2.28). The solid curve drawn through the dots is merely a guide to the eyes. The dashed line is the phase boundary in the absence of hyperfine interaction, as shown also in Fig. 2.1. The inset shows the experimentally obtained phase boundary of Bitko et al. from susceptibility measurements (filled circles) and the calculated mean-field phase boundary (solid line) in the presence of hyperfine coupling from an 136 x 136 eigenfunction space of the full Hamiltonian (nuclear spin = 7/2, ionic spin = 8) [48]. The dashed line in the inset is the result of computation in the absence of hyperfine interactions.

The deviation of the phase boundary from the dashed line for high values of the transverse field Γ can be understood from the following physical considerations. For small Γ the local mean-field is dominated by the effective Ising interaction proportional to the magnetization along the c-axis. However, as Γ increases, the magnetization along the c-axis goes on decreasing (because of the gradual loss of the quantization axis) leading to the situation in which the local field is totally dominated by the hyperfine coupling alone. Because it is the competition between the local field and the transverse field that determines the phase boundary between the ferromagnetic and paramagnetic phases, the ferromagnetic order persists for a larger values of the transverse field, especially at low temperatures, since the hyperfine field takes over the role of the local mean field. This analysis is in conformity with the experimental observation of Bitko et al. in that there is an upward tilt in the phase boundary for temperatures lower than 0.6K [48].

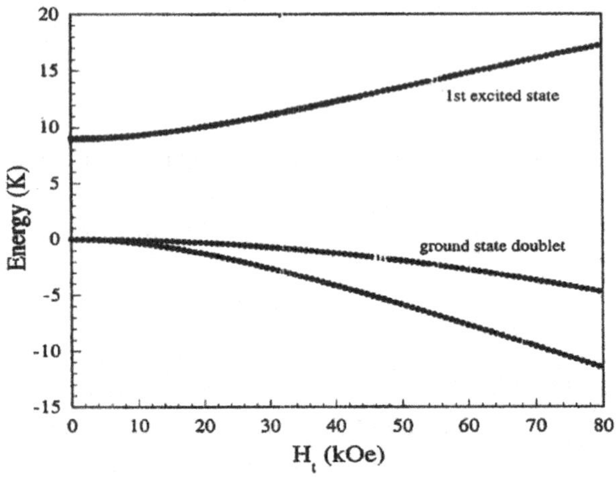

Fig. 2.3. The crystal field splitting in $LiHoF_4$, indicating that at ultra low temperatures, one has an effective 2-state model.

2.4. Quantum Statistical Mechanics

With the preceding background to quantum phase transition (QPT) in the context of an example borrowed from Magnetism we turn now to the general statistical mechanics discussion on why a QPT is so different from its classical counterparts. Note that the main issue in evaluating the partition function in statistical mechanics concerns the diagonalization of the Hamiltonian. Once this is done one can choose to work in a representation in which the Hamiltonian is diagonal and then compute the trace of the density operator $e^{-\beta\mathcal{H}}$ merely in terms of the eigenvalues of $\mathcal{H}$, for completing the calculation of the partition function. The reason this task is difficult in general in a quantum problem is evident from the example of the transverse Ising model; the Ising term is diagonal in the representation of the spin operator S_z, while the transverse term is diagonal in the representation of the spin operator S_x ! Thus there is no definite axis of quantization and hence, the task of diagonalization is humongous, barring the special case of one dimension, or finite systems. This problem is of course peculiar to quantum mechanics because in classical statistical mechanics the kinetic energy and potential energy terms (which add up to the Hamiltonian $\mathcal{H}$) can be dealt with separately. The term in the partition function involving the kinetic energy yields a Gaussian integral over the momentum variable that leads to a simple temperature – dependent prefactor and all the many body effects are subsumed in the potential energy term. Not that it always leads to a happy situation as far as an exact calculation of the partition function is concerned, as we very well know from the examples of the ordinary Ising model in three dimensions, or interacting classical liquids, etc., but the first battle is won, viz., the kinetic and potential energy terms can be disentangled. Indeed it is this attempt towards disentanglement that yields the basis of the Feynman's path integral formulation of the partition function, as discussed below [53].

The partition function in the canonical ensemble can be written in terms of a summation over a complete set of states:

$$Z(\beta) = \sum_f < f|e^{-\beta\mathcal{H}}|f >. \tag{2.29}$$

Note that the density operator $e^{-\beta\mathcal{H}}$ is identical to the time-development operator $e^{-i\mathcal{H}\tau/\hbar}$, if the time τ is considered to be imaginary: $\tau = -it\beta$. Thus Z is the sum of transition amplitudes for the system to start from some state $|f>$ and return to the same state $|f>$ after a time interval $-i\hbar\beta$. The total time interval, which is fixed by temperature, will naturally be infinitely long (along the imaginary axis) at $T = 0$!

The idea of Feynman is to break up the net transition amplitude between the states of a system into a series of transition amplitudes over intermediate states and summing over all possible paths between the two states. The path in question is defined by specifying the state of the system as a sequence of finely spaced intermediate time steps. One may formally write

$$e^{-\beta\mathcal{H}} = [e^{-\frac{1}{\hbar}\delta\tau\mathcal{H}}]^N, \tag{2.30}$$

where $\delta\tau$ is a small time interval given by $\delta\tau = \hbar/_E$ where E is taken to be larger than the highest energy scale in the problem, and N is an integer chosen such that $N\delta\tau = \hbar\beta$. We now use the Trotter identify [54]:

$$exp(\hat{A} + \hat{B}) = \lim_{N\to\infty}(e^{\frac{\hat{A}}{N}}\ e^{\frac{\hat{B}}{N}})^N, \tag{2.31}$$

where $\hat{A}$ and $\hat{B}$ are two bounded quantum operators.

Writing $\mathcal{H}$ as $\mathcal{H} = \mathrm{T} + V$, where T is the kinetic energy operator and V the potential energy operator, the kinetic and potential energy terms can be disentangled in the limit $\delta\tau \to 0$. Hence the idea of employing a small time interval $\delta\tau$ is to enable the use of classical-like methods by ignoring the non-commulativity of T and V over the small time segment $\delta\tau$! Using a succession of closure property the partition function may be rewritten as

$$Z(\beta) = \sum_f \sum_{f_1 f_2 \cdots f_N} < f \left| e^{-\frac{1}{\hbar}\delta\tau H} \right| f_1 >$$

$$\mathrm{x} < f_1 | e^{-\frac{1}{\hbar}\delta\tau\mathcal{H}} | f_2 >< f_2 | \ldots | f_N >< f_N | e^{-\frac{1}{\hbar}\delta\tau\mathcal{H}} | f >. \tag{2.32}$$

Note that a typical matrix element such as $< f_1 \left| e^{-\frac{1}{\hbar}\delta\tau H} \right| f_2 >$ signifies an amplitude of transition (in imaginarily time $i\delta\tau$) from the state $|f_1>$ to the state $|f_2>$. This is reminiscent of a transfer matrix that connects one site to another in classical statistical mechanics of the Ising model, for instance [17]. Therefore, in the present case, the imaginary time may be thought of as an extra spatial dimension. This indeed is the reason why the statistical mechanics of a d-dimensional quantum system can be viewed equivalent to a $(d+1)-$ dimensional classical system, except that the extra dimension is finite in extent, going to infinity only at $T=0$. The underlying idea can be illustrated in terms of the following physical example.

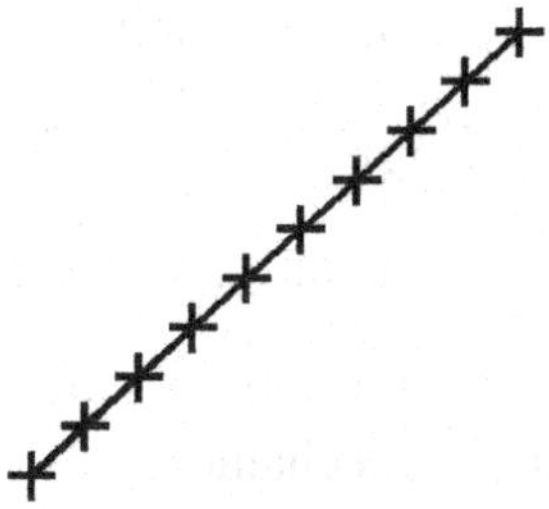

Fig. 2.4. Josephson junctions (crosses) demarcating superconducting grains.

Consider a one-dimensional array of Josephson junctions shown schematically in Fig. 2.4 [44]. The crosses indicate junctions between superconducting grains. Over each grain the superconducting state is characterized by a condensate wave function. Josephson tunneling of cooper pairs from one grain to another occurs across the junctions. The order parameter at hand is the condensate wave junction itself which for the jth grain can be written as $\psi_j = a_j e^{i\theta_j}$. As it turns out it is the fluctuations of the phase angle θ_j that are relevant while those of the amplitude a_j can be ignored. Because tunneling involves overlap of the wave function between grains, the corresponding interaction Hamiltonian is proportional to the projection operator $|\psi_j><\psi_{j+1}|$ between the jth and its neighboring $(j+1)^{th}$ grains. If coherence of the wave function is maintained, tunneling is facilitated and the system (i.e. the one-dimensional array) is super-conducting. This process however is counter balanced by the energy required to transport an excess cooper pair onto a grain, i.e. the coulomb repulsions between pairs. If this energy is large, especially for small grains, the Cooper pairs will be blockaded and the system becomes an insulator.

It is this superconductor-insulator transition which is the underlying QPT in the present instance. Given this background the Hamiltonian for the Josephson junction array can be written as:

$$\mathcal{H} = \frac{1}{2} C \sum_j (\hat{V}_j)^2 - \sum_j E_j \cos(\hat{\theta}_j - \hat{\theta}_{j+1}), \tag{2.33}$$

where the hat over a dynamical variable the phase angle signifies its operator nature, E_j is the Josephson coupling strength, C is the intra-grain capacitance and $\hat{V}_j$ is the 'voltage operator' for the jth grain. Because $\hat{V}_j$ is conjugate to the dynamical variable $\hat{\theta}_j$ it has a differential operator form (analogous to the co-ordinate representation of the momentum operator in quantum mechanics):

$$\hat{V}_j = -i\left(\frac{2e}{C}\right)\frac{\partial}{\partial\theta_j}. \tag{2.34}$$

(Note that the extra factor of 2 in front of e accounts for the total electric charge of a Cooper pair). Substituting Eq. (2.34) in Eq. (2.33) the Hamiltonian can be rewritten as

$$\mathcal{H} = -E_c \sum_j \frac{\partial^2}{\partial\theta_j^2} - \sum_j E_j \cos(\hat{\theta}_j - \hat{\theta}_{j-1}), \tag{2.35}$$

where

$$E_c = \frac{2e^2}{C}. \tag{2.36}$$

The first term in Eq. (2.35) is analogous to the kinetic energy of a rotor where E_c^{-1} is like the moment of inertia.

At this stage it is possible to draw an analogy between the Hamiltonian in Eq. (2.35) and that of the TIM (cf., Eq. (2.1)). The Josephson energy is akin to the pure Ising coupling in that each interaction favors 'order' be it in the form of preserving the phase coherence of the condensate wave function or ferromagnetism respectively. The charging energy, proportional to E_c is like the transverse term Γ of TIM that has the effect of disrupting order by effecting quantum transitions between the eigenstates of the angle operator $\hat{\theta}_j$ (or the spin operator S_{jz}, in the TIM case). Understandably then, the disordering field or the control parameter for the quantum phase transition in the Josephson junction array is expected to be proportional to the ratio of E_c and E_j. (For a precise relation, see below). When this ratio exceeds a certain critical value the system transits from the superconducting to the insulating phase.

As discussed before in the context of the expression for the partition function (Eq. 2.32), each transition amplitude $< f_1 | e^{-\frac{1}{\hbar}\delta\tau\mathcal{H}} | f_2 >$, which is a statistical weight of a space time path, is the Boltzmann weight of the classical configuration for an equivalent 2d-classical model, in which the extra dimension is along the imaginary time axis. The latter turns out to be the 2-dimensional classical XY-model, the Hamiltonian of which has already been introduced in Eq. (1.3). The angles θ_i are now to be interpreted as those of planar spins that lie on a two-dimensional square lattice. It turns out that the coupling constant J is proportional to $\sqrt{E_J / E_C}$. The inverse of J is then a 'fake' temperature for the model, the increase of which causes the coherent ordering of the phases (for the original model) to disappear. On the other hand, a large value of J makes it unlikely that the angles θ_i and θ_j will differ significantly, even when the sites i and *j are far apart (in space and/*or time), yielding a superconducting phase. We should emphasize that this equivalence between the one-dimensional Josephson junction array and the two-dimensional classical XY-model is only in the sense of universality near the corresponding critical points (cf., Chapter I). Other than that however the underlying proof of equivalence goes over to d-dimensional arrays and $(d+1)-$dimensional classical XY models. This is a very useful result in that numerical methods such as Monte Carlo techniques can be applied to study the critical behaviour of d-dimensional quantum models.

In conclusion, we note that all the basic ideas of quantum phase transitions and the concomitant quantum critical points have not only a historical connotation to Magnetism but are also most easily understood within the context of a magnetic model eg. the transverse Ising model. Further details of zero temperature $(T=0)$ dynamic scaling, finite-size scaling for $T \neq 0$, and Quantum Hall Effect as a prototypical experimental system of QPT can be found in the eminently readable review article of Sondhi et al. [44].

Chapter 3

Glass Transitions

3.1. Preamble – Magnetic Glass, Quantum Glass, Spin Glass, Proton Glass and Structural Glass

One of the enigmatic unresolved issues of the contemporary materials physics is the glass transition. The problem is simply stated in the context of an ordinary, monatomic liquid. When the latter is cooled very slowly, i.e. adiabatically such that at every stage of the cooling process the liquid is in thermal equilibrium with the heat bath in which it is embedded, it usually freezes into a crystalline solid at a temperature T_F. However, when the liquid is rapidly cooled, i.e. supercooled, it solidifies into a disordered amorphous structure called glass, which has no translational or orientational symmetry unlike a crystal. The transition temperature T_G ($<T_F$) is an experimental one, that is conventionally taken as the temperature at which the viscosity reaches a value of 10^{13} Poise. The dynamical behavior of the metastable supercooled state has many unusual features such as non-exponential, multi-stage time decay of fluctuations and a rapid growth of relaxation times, associated with the increase in viscosity [55–60].

While there is still no consensus about a satisfactory theoretical scenario for the observed glass transitions many of the experimental signatures are shared by completely different systems that were historically investigated in Magnetism. Inevitably also, much of the novel theoretical concepts, required for the analysis of the glass transition, such as a new order parameter, replica technique for calculating the partition function, non-ergodic phase space dynamics, etc., arose also in the context of these magnetic systems [61–65]. Therefore, in Sec 3.1, we discuss one of these magnetic paradigms called a Magnetic Glass, which is simply a cousin of our earlier studied system of $LiHoF_4$ in which the rare earth Ho is randomly replaced by another rare-earth element Y (Yittrium) which is nonmagnetic, however [66]. The system of $Li(Ho)_x\ Y_{1-x}F_4$ in a magnetic field, applied transverse to the c-axis, provides a disordered version of the TIM, discussed in Chapter 2, and is logically referred to as a Quantum Glass, which is the topic of Sec. 3.3. While the Magnetic Glass and the Quantum Glass are based on the Ising model, the first evidence of glass-like behavior in a solid state system is known to occur in a dilute magnetic alloy of a nonmagnetic metal, e.g., Mn in Cu, called a Spin Glass, which is described by a classical Heisenberg model, introduced earlier in Chapter 1 [67]. Spin Glasses are briefly discussed in Sec. 3.4. In Sec. 3.5 we return to the Magnetic Glass and present an analogous system named Proton Glass which is a random admixture of the ferroelectric crystal of $(Rb)H_2\ PO_4$ and the antiferroelectric crystal of $(NH_4)\ H_2\ PO_4$ [68]. Once again, it is interesting to note how ideas developed in the magnetic milieu find their echo in the vastly different context of a structural glass, which forms our brief concluding remarks of this chapter in Sec. 3.6.

3.2. Magnetic Glass: Li $Ho_x\ Y_{1-x}\ F_4$

We come back to the rare earth magnet of Li Ho F_4 discussed in Sec. 2.3. Recall that it is the holmium ion that carries a magnetic moment. The relevant many body interaction is then provided by the dipolar coupling between the moments on different holmium ions in the underlying insulating host matrix (See Fig. 2.2). We now consider a situation in which a non-moment carrying but a similar rare earth atom of Yttrium is randomly introduced at holmium sites. This amounts to breaking of bonds randomly for the underlying Ising model for which the bond strength is described by Eq. (2.22). We are then led to consider a situation in which the Ising Hamiltonian described by

$$\mathcal{H} = -\sum_{ij} J_{ij}\, S_{iz} S_{jz}, \qquad (3.1)$$

is characterized by random bond strengths which must be specified with the aid of a probability distribution.

As mentioned in Chapter 2, even without the introduction of the impurity atom of Yttrium the ordered crystal of $LiHoF_4$ is endowed with competing Ising interactions, with the bonds switching from ferromagnetic to antiferromagnetic types, depending on the angle θ_{ij} (cf., Eq. (2.22)). The antiferromagnetic case is special, and it merits a separate discussion, even for a regular crystalline lattice. Consider then, for the sake of simplicity of discussion, a two-dimensional triangular lattice shown in Fig. 3.1. We focus on the triangle ABC and insert a

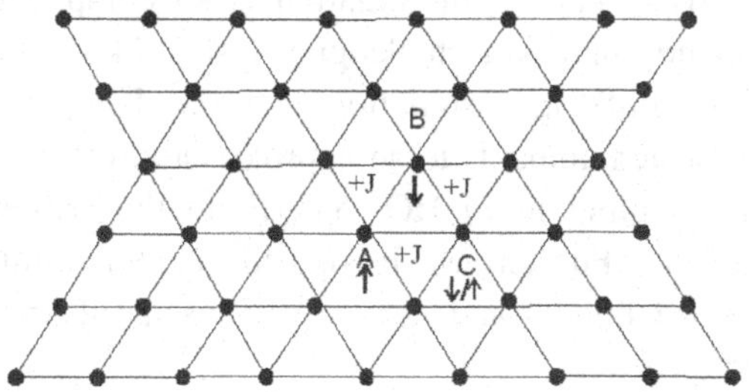

Fig. 3.1. A triangular lattice with antiferromagnetic coupling, illustrating Frustration.

nearest neighbor bond of strength +J between each pair of vertices on which reside two spins. Assuming that we start from the vertex A on which the spin is 'up', the B-spin would like to be 'down', in order to minimize antiferromagnetic energy. Keeping in view the orientation of the B-spin, the C-spin would like to be 'up' but from the point of view of the A-spin, it would like to be 'down'. Thus, there are conflicting 'messages' at the C-site from the nearest neighbours as a result of which the C-spin is 'frustrated'. Frustration implies that the ground state of the underlying Hamiltonian is not unique – it is degenerate. We should reiterate, by way of this example, that frustration can occur even for a regular crystalline solid, if we recall that a triangular lattice is just a two-dimensional projection of a body-centered cubic (bcc) crystal.

What then is the situation in $LiHoF_4$, which is endowed with both ferromagnetic and antiferromagnetic bonds and the underlying lattice is bcc? It is here that the ingenious analysis of Luttinger and Tisza assumes significance [46]. These authors carefully examined the dipolar interaction for a bcc system and proved that the ground state is ferromagnetic – there is no frustration. However, in general, if there is an admixture of ferro and anti-ferromagnetic bonds, frustration is likely to occur, as illustrated schematically in Fig. 3.2. We show a single square of Ising spins for which the nearest neighbor coupling can be both ferromagnetic (– J) and antiferromagnetic (+ J). Again, for the spin configurations shown in the diagram the spin at the right bottom-corner is frustrated as it does not know which way to point, in order to minimize energy. Thus, when both frustration and disorder are simultaneously present as in LiHo$_x$ Y$_{1-x}$ F$_4$ the situation is so complicated that even finding the ground state is a challenging task. In view of the above we make the simplifying assumption that the Ising interactions are infinite ranged, which amounts to an approximate treatment of the long-ranged dipolar coupling in Li HO$_x$ Y$_{1-x}$ F$_4$ in the presence of strong crystal field terms. Further, we invoke the central limit theorem for a large number (of bonds) and assume a Gaussian distribution for the bond strengths:

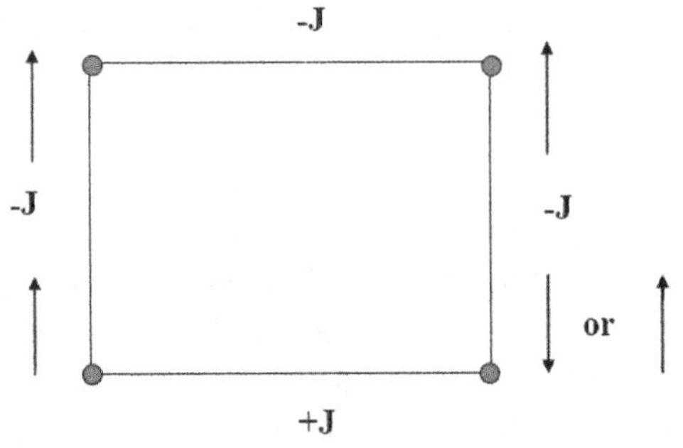

Fig. 3.2. Ising spins on a square lattice with both ferro – and antiferromagnetic bonds, illustrating the idea of frustration for a random system.

$$P(J_{ij}) = \frac{1}{\sqrt{2\pi\Delta^2}} exp\left[-\frac{(J_{ij} - J_o)^2}{2\Delta^2}\right]. \tag{3.2}$$

Because the concentration of magnetic moment carrying elements of Ho is x, for which the coupling can be either ferromagnetic (– J) or antiferromagnetic (+ J) we assume + J and – J to occur with equal likelihood and therefore be associated each with concentration $^x/_2$. On the other hand, nonmagnetic Y corresponds to zero bond strength, occurring with concentration (1-x). Hence, the mean bond strength can be estimated as

$$J_o = \frac{x}{2}(J) + \frac{x}{2}(-J) + (1-x).0 = 0. \tag{3.3}$$

On the other hand, the variance is

$$\Delta^2 = \frac{x}{2}(J^2) + \frac{x}{2}(J^2) + (1-x).0 = xJ^2. \tag{3.4}$$

Hence, Eq. (3.2) reduces to

$$P(J_{ij}) = \frac{1}{J\sqrt{2\pi x}} exp\left(-\frac{J_{ij}^2}{2xJ^2}\right). \tag{3.5}$$

An infinite-ranged Ising model with a Gaussian probability distribution as in Eq. (3.5) goes under the name of the Sherrington-Kirkpatrick model [69, 70].

The occurrence of frustration and disorder necessitates a new kind of statistical mechanical treatment. Recall that in the canonical ensemble of statistical mechanics, ensemble averaging of dynamical variables is carried out with the aid of a partition function defined by

$$Z = Tr\ (exp(-\beta\mathcal{H})). \tag{3.6}$$

Thus the magnetization m is proportional to the ensemble average (denoted by angular brackets) of the spin, given by

$$< S_{jz} >= \frac{1}{Z} Tr[S_{jz} exp\ (-\beta\mathcal{H}|]. \tag{3.7}$$

The reason that m can be directly compared with experiments is that all possible configurations (=2^N, for an N-spin Ising system) associated with the Hamiltonian $\mathcal{H}$ are assumed to have been realized through phase space exploration within the measurement time. Thus the partition function Z provides a link between the 'microscopic' subject of statistical mechanics with the 'macroscopic' subject of thermodynamics through the relation.

$$\mathcal{Z} = exp(-\beta F), \tag{3.8}$$

where F is the Helmholtz Free Energy.

On the other hand, when we are confronted with disorder due to 'quenched' dynamical variables, e.g. Y in place of H_o, these variables are in a sense 'frozen' and are therefore outside the scope of reckoning as far as usual, statistical averaging is concerned. Therefore, the Free energy that is observationally relevant is to be calculated from the 'thermodynamic' partition function (or more correctly, its logarithm) by performing an additional 'quenched averaging', denoted by a bar below. Thus

$$F = -k_B T\, \overline{ln\, \mathcal{Z}} \tag{3.9}$$

Correspondingly, the measured magnetization is related to:

$$\overline{m}_j = \overline{< S_{jz} >}, \tag{3.10}$$

where the angular brackets signify usual thermodynamic averaging in the sense of Eq. (3.7) and the bar specifies averaging over the quenched disorder. In other words,

$$\overline{m}_j = \frac{\int d(J_{ij})P(J_{ij})\, Tr\,(S_{jz}e^{-\beta\mathcal{H}(\{J_{ij}\})})}{\int d(J_{ij})P(J_{ij})\, Tr\,(Se^{-\beta\mathcal{H}(\{J_{ij}\})})}. \tag{3.11}$$

It is evident that in the case described above in which $J_o = 0$ because of symmetric distribution of $\pm J$ bonds, $\overline{m} = 0$. The result spurred Edwards and Anderson to introduce a new order parameter defined by [71]:

$$q = \overline{m_j^2} = \overline{< S_{jz} >^2}. \tag{3.12}$$

Note that $< S_{jz} >$ is calculated in the usual thermodynamic sense for a given fixed value of J and is therefore, nonzero, except in the paramagnetic phase. The averaging over the random distribution of J_{ij} (indicated by the overhead bar) is done only after squaring $< S_{jz} >$. Evidently, q remains zero in the paramagnetic phase and hence the occurrence of nonzero q indicates an unusual ordering phenomenon yielding what is called a spin glass phase. Note that after the quenched averaging is performed, q becomes independent of the site index j.

Following extensive experimental and theoretical investigations it has become apparent that in analogy with q one needs a different response function or susceptibility χ_{SG} which is expected to diverge as $T \to T_{SG}^{+}$, where T_{SG} is the temperature at which the spins freeze in some random-looking orientation. The spin glass susceptibility is then defined by

$$\chi_{SG} = \frac{1}{N}\sum_{i,j} \overline{< S_{ij} S_{jz} >^2}. \tag{3.13}$$

It is evident that χ_{SG} is accessible in computer simulations. Further, when it comes to laboratory experiments, χ_{SG} is essentially the same as the <u>nonlinear</u> susceptibility which can be read off as the coefficient of H^3 in the expansion of the magnetization in the powers of the applied field H:

$$m = \chi H - \chi_{NL} H^3 \tag{3.14}$$

As $T \to T_{SG}^{+}$, χ_{NL} diverges as

$$\chi_{NL} \sim (T - T_{SG})^{-\gamma}. \tag{3.15}$$

where τ is the usual critical point exponent.

The spin glass transition seen in magnetic systems has motivated the use of not just an unusual order parameter q and its associated susceptibility χ_{SG} for other glassy systems as well, but has also given rise to new methods in Statistical Mechanics. Thus, after employing a certain representation of the logarithm of the partition function, $\mathcal{Z}$, the Free Energy in Eq. (3.9) can be written as

$$F = -k_B T \lim_{n \to 0} \overline{[(\mathcal{Z}^n - 1)/n]}. \tag{3.16}$$

Note that it is not $\mathcal{Z}$ but ln$\mathcal{Z}$ which has to be averaged because of its connection to the extensive variable: the Free Energy. It is ln$\mathcal{Z}$ that is self-averaging.

The underlying idea of the <u>Replica Method</u> is to imagine that for each realization of the bond strength J_{ij} there is a copy or replica of the system [71]. The trick is then to read the nth power of Z in conjunction with the overhead bar as the n-times product of the partition function for a given replica. Thus

$$\overline{Z^n} = \overline{(Tr\, e^{-\beta\mathcal{H}})^n Tr} = \overline{Tr\, e^{-\beta\sum_{\alpha=1}^{n}\mathcal{H}_\alpha}} = \prod_{\alpha=1}^{n}\overline{(Tr_\alpha e^{-\beta\mathcal{H}_\alpha})} = \prod_{\alpha=1}^{n}\overline{(Z_\alpha)}.$$

(3.17)

Thus we have 'replicated' the system n times. For the symmetric Gaussian bond distribution ($J_0 = 0$) in Eq. (3.5), we obtain

$$\overline{Z^n} = Tr_{S_1\ldots S_n}\, exp\left(\frac{1}{4}\beta^2\Delta\sum_{i,j}\sum_{\gamma\beta} S_{iz}^{\alpha}\, S_{iz}^{\beta}\; S_{jz}^{\alpha}\; S_{jz}^{\beta}\right), \tag{3.18}$$

where the index attached to the spin is called a replica index. Hence the disordered problem is converted into a non-random one, but with a 4-spin interaction Hamiltonian.

Accordingly, the Edwards–Anderson order parameter q can be rewritten as

$$q = \lim_{n\to 0}\frac{1}{N}\sum_{j=1}^{N}\sum_{\alpha,\beta=1}^{n} < S_{jz}^{\alpha}\, S_{jz}^{\beta} >, \tag{3.19}$$

where α, β correspond to different replicas.

With the preceding background we can now discuss the order parameter relation for the magnetic glass system of $LiHo_xY_{1-x}F_4$ in the mean field approximation. Recall in the latter the single spin Hamiltonian can be written as

$$\mathcal{H}_i = -H_i\, S_{zi}, \tag{3.20}$$

where

$$H_i = \sum_{j=1}^{N} J_{ij} < S_{zj} > \equiv \sum_{j=1}^{N} J_{ij} m_j. \tag{3.21}$$

In accordance with Eq. (3.5), the effective local fields H_i can also be assumed to be Gaussian-distributed as follows:

$$P(H_i) = \frac{1}{\sqrt{2\pi\overline{H^2}}} exp\left(-\frac{H_i^2}{2\overline{H^2}}\right), \tag{3.22}$$

where $\overline{H^2}$ is the appropriate variance, given by

$$\overline{H^2} = \overline{(\sum_{j=1}^{N} J_{ij} m_j)^2}. \tag{3.23}$$

Assuming that J_{ij} and m_j are independently distributed [72],

$$\overline{H^2} = \sum_{j=1}^{N} \overline{J_{ij}^2}\ \overline{m_j^2} = J^2\overline{m_j^2}\,. \tag{3.24}$$

On the other hand, m_j is given by (cf., Eq. (2.10)) as

$$m_j = tanh(\beta H_j). \tag{3.25}$$

Identifying $\overline{m_j^2}$ in Eq. (3.24) with the Edwards–Anderson order parameter q and employing Eq. (3.22), we obtain the self-consistent equation.

$$q = \frac{1}{\sqrt{2\pi\overline{H^2}}} \int_{\infty}^{\infty} dH (tanh\beta H)^2 exp\left(-\frac{H^2}{2\overline{H^2}}\right). \tag{3.26}$$

We introduce now a dimensionless variable ξ as

$$\xi = \frac{H}{J\sqrt{q}}, \tag{3.27}$$

in terms of which

$$q = \frac{1}{\sqrt{2\pi}} \int_{-\infty}^{\infty} d\xi\ (tanh\,\beta H(\xi))^2 e^{-\xi^2/2} \tag{3.28}$$

$$= \frac{1}{\sqrt{2\pi}} \int_{-\infty}^{\infty} d\xi p(\xi)^2 e^{-\xi^2/2}, \tag{3.29}$$

where $p(\xi)$ is the local polarization given by

$$p(\xi) = tanh\beta H(\xi). \tag{3.30}$$

The steps needed for evaluating physically relevant quantities, e.g., the partition function, the free energy, correlation functions, etc., are as stated below. We consider a given realization of the disorder, e.g., a prescribed value of the bond strength J_{ij} as in the present case, and then perform an average over the underlying probability distribution of the disorder. In Mean Field Theory the disorder effects are entirely encapsulated within the polarization p. Thus, all physical quantities are first calculated for a fixed value of p, and the results are then averaged over the distribution of p, denoted by $W(p)$, defined by

$$W(p) = \frac{1}{N}\sum_{j=1}^{N} \delta\left(p - < S_{zj} >\right) = \int d\xi\ \delta[p(\xi) - p], \tag{3.31}$$

where the last term is written as a functional integral. Denoting the argument of the delta function as $f(\xi)$, where

$$f(\xi) = tanh[\beta H(\xi)] - p, \tag{3.32}$$

and recalling the identity

$$\delta\,[f(\xi)] = \frac{\delta(\xi - \xi_0)}{f'(\xi_0)}, \tag{3.33}$$

where $f_o(p)$ is the solution of $f\ [\xi_o(p)] = 0,$ we obtain

$$W(p) = \frac{1}{\sqrt{2\pi}}\int_{-\infty}^{\infty} d\xi e^{-\xi^2/2}\delta[f(\xi)] = \frac{1}{\sqrt{2\pi}}\frac{e^{-\xi_0{}^2/2}}{f'(\xi_0)}. \tag{3.34}$$

It is evident from Eq. (3.32) that

$$f'(\xi) = \beta J\sqrt{q}[1 - p(\xi)^2]. \tag{3.35}$$

Hence

$$W(p) = \frac{1}{\sqrt{2\pi q}}\frac{e^{-\xi_0^2(p)/2}}{\beta J}(1 - p^2)^{-1}. \tag{3.36}$$

The steps leading to the above result are fraught by several assumptions which are invalid in essence. For instance, the fields at different sites are correlated due to disorder, even within the mean field approximation, and hence the central-limit theorem, implicitly assumed in Eq. (3.22), is not applicable. In addition, the site magnetizations are not independent of each other or of the J_{ij}'s hence the factorization assumed in Eq. (3.24) is incorrect. Finally, even the mean field equations are invalid for a magnetic glass even if they are reasonable, within limitations, for the corresponding ferromagnet. Yet, and rather mysteriously, the above results are in agreement with those derived from more rigorous treatments, e.g., replica-symmetric theory or the Langevin dynamical formulation of the long-ranged mean field models [72]. To summarize, the results obtained in the crude mean field theory, outlined above, are qualitatively reasonable, although their derivation is wrong in details, above the so called Almeida-Thouless line in the phase diagram [73]. Below that line, the assumption of replica symmetry breaks down and so does the above mean field theory as the physics is considerably more intricate.

3.3. Quantum Glass: Li Ho_x Y_{1-x} F_4 in a Transverse Field

We have earlier discussed in Chapter 2 the transverse Ising model via the example of $LiHoF_4$ – system. We now consider the doped system of $LiHo_xY_{1-x}F_4$, but in a magnetic field transverse to the c-axis. The result is the same TIM Hamiltonian given in Eq. (2.1) but now the bonds J_{ij} are randomly distributed with a probability given by Eq. (3.2). The glassiness arising from the random substitution of Ho by Y already causes a frustrated ground state and yields a free energy landscape with multiple minima. The presence of the transverse field can lead to additional quantum tunneling processes connecting these minima. The question that naturally arises is: Do quantum interactions preserve or destroy phase transitions in these glassy system? The doped system of $LiHo_xY_{1-x}F_4$ in a transverse field provides the 'World's Simplest 'Quantum Glass' to address such questions, again within the paradigm of Magnetism.

The analysis of the Edwards-Anderson order parameter simply follows the mean field treatment of the Magnetic Glass, outlined in Sec. 3.2 above. (cf., Eq. (3.20) - (3.36)). Accordingly, Eq. (3.20) is modified as

$$\mathcal{H}_i = -h_i\, S_{zi}, \tag{3.37}$$

where

$$h_i = \sqrt{H_i^2 + \Gamma^2}. \tag{3.38}$$

Correspondingly, Eq. (3.25) reads

$$m_j = \frac{H_j}{h_j} \tan h(\beta h_j), \tag{3.39}$$

and the order parameter equation (3.26) is generalized to:

$$q = \frac{1}{\sqrt{2\pi}} \int_{-\infty}^{\infty} d\xi \left[\frac{H(\xi)}{h(\xi)} \tanh\,(\beta h(\xi))\right]^2 e^{-\xi^2/2}. \tag{3.40}$$

Finally, the function $f(\xi)$ that occurs in the argument of the delta function, under the integral for the probability distribution W(p) for the polarization (cf., Eq. (3.34)), can be rewritten

$$f(\xi) = \frac{H(\xi)}{\sqrt{H(\xi)^2 + \Gamma^2}} \tanh\left(\beta\sqrt{H(\xi)^2 + \Gamma^2}\right) - p. \tag{3.41}$$

Hence

$$W(p) = \frac{e^{-\xi_0{}^2/2}}{\sqrt{2\pi q}} \frac{1}{\beta J} \left[\frac{H(\xi_0)^2}{H(\xi_0)^2 + \Gamma^2} + \frac{\Gamma^2}{\beta} \frac{p(\xi_0)}{H(\xi_0)(H(\xi_0)^2 + \Gamma^2)} - p(\xi_0)^2\right]^{-1}, \tag{3.42}$$

where ξ_0, as before, equals $\xi_0(p)$ which is the inverse function of $p(\xi)$ that satisfies.

$$p(\xi) = \frac{H(\xi)}{\sqrt{H(\xi)^2 + \Gamma^2}} \tanh\left(\beta\sqrt{H(\xi)^2 + \Gamma^2}\right). \tag{3.43}$$

Needless to say, the above mean field equations suffer from the same criticisms as the earlier discussed equations for the magnetic glass but again, quite remarkably, these simple-minded equations are validated by a more elaborate thermo-field dynamics approach [74], especially above the Almeida-Thouless – like surface. The mean field theory, outlined above, forms the basis of analyzing the so-called Non-Debye behavior of the measured frequency-dependent susceptibility in the Magnetic Glass [75].

3.4. Spin Glasses

As mentioned in the beginning of this chapter the first glassy system in Magnetism that was extensively studied in the laboratory in the 1970's is what is called a Spin Glass. A Spin Glass, unlike the magnetic glass in which the relevant interaction is a dipolar one in an insulating host, is a metallic system e.g. Cu, in which magnetic moment carrying elements, e.g. Mn are quenched-in a rapidly cooled solid solution. The resulting alloy in which the Mn concentration is low is disordered because of random substitution of the host Cu atom by the magnetic impurity of Mn. The spin on the Mn site is essentially localized but it strongly interacts with the s-wave itinerant electron of the host Cu metal via an exchange coupling. The itinerant electrons themselves are in long-ranged communication with each other in which the range of the interaction is governed by the inverse of the Fermi wave vector k_F. Thus, mediated by the itinerant electrons of the host metal, there occurs an indirect coupling between the localized moments of Cu which is referred to as the Ruderman–Kittel–Kasuya–Yoshida (RKKY) interaction. This coupling can be described by the classical Heisenberg model, the Hamiltonian of which was introduced in Chapter I, but now the bond strength is governed by

$$J_{ij} = J_0 \frac{cos(2\overrightarrow{k_F}\,.\overrightarrow{R_{ij}})}{\left|\overrightarrow{R_{ij}}\right|^3}, \tag{3.44}$$

where J_o is a constant. Because the cosine term fluctuates in sign the interaction switches from ferromagnetic to antiferromagnetic leading to frustration. Besides, the vector distance $\overrightarrow{R_{ij}}$ between two Mn moments is randomly distributed because of the manner in which the $Cu(Mn)_x$ solid solution is prepared. Thus the underlying system has both the important attributes of a magnetic glass, viz., frustration and disorder, and like the dipolar–coupled system of Sec. 2.2, is also endowed with a long-ranged interaction that falls off as $\frac{1}{R^3}$. However, the symmetry of the interaction is now Heisenberg–like and not Ising–like. This is because Mn is an S-state ion and therefore has very little anisotropy.

All the underlying novel concepts associated with glassy systems such as the Edwards–Anderson order parameter, replica methods, replica– symmetry, etc. were historically introduced in the context of the canonical spin glasses e.g. $Cu(Mn)_x$, $Au(Fe)_x$. There has been a plethora of experimental studies in spin glasses which have established that a spin glass has a sharp thermodynamic phase transition at a temperature $T = T_{SG}$, such that for $T < T_{SG}$ the spins freeze in random orientations, reflecting the vector nature of the coupling. Experiments also reveal a divergence of the spin glass susceptibility which is now defined by Eq. (3.13).

While the existence of a phase transition is not an issue, atleast for three dimensions, the nature of the equilibrium state below T_{SG} is much more controversial. This is because experimentally, it is impossible to access the equilibrium state below T_{SG} in view of very long relaxation times and the system getting trapped (almost forever) in a metastable valley.

Although our entire discussion on Spin Glasses has been focused on static effects and time-dependent phenomena have been relegated to Chapter 4 onwards, the understanding of unusual spin glass ordering remains incomplete without invoking dynamical concepts. This point is elaborated with the aid of a schematic illustration in Fig. (3.3). What is shown is the nature of spin orientations in an ordered ferromagnet, a paramagnet and a spin glass at times $t = 0$ and $t = \infty$. From the diagram it is clear that the phase transition in a spin glass is associated, not with a long range order in space, but with a kind of order in time. There exists a sort of 'memory effect', to be discussed in far more detail in Chapter V, linked with the time fluctuations of a given spin. Consequently, a spin glass exhibits subtle hysteresis and other phenomena which are directly connected with the 'measurement-time' of a particular experiment [76]. Thus non-equilibrium effects, often referred to as <u>non-ergodicity</u> in the context of glasses, are of fundamental importance in a random system. It is therefore natural to introduce a time-dependent quantity which describes the configurationally (i.e. 'quenched') averaged auto-correlation of spin moments of a spin glass [77]:

$$q(t) = \frac{1}{N}\sum_{i=1}^{N} \overline{<\vec{S_i}(0).\vec{S_i}(t)>}, \qquad (3.45)$$

where, as used earlier, the angular brackets $< \cdots >$ denote the usual thermal average, while the bar implies an additional configuration average over the quenched variables of the system.

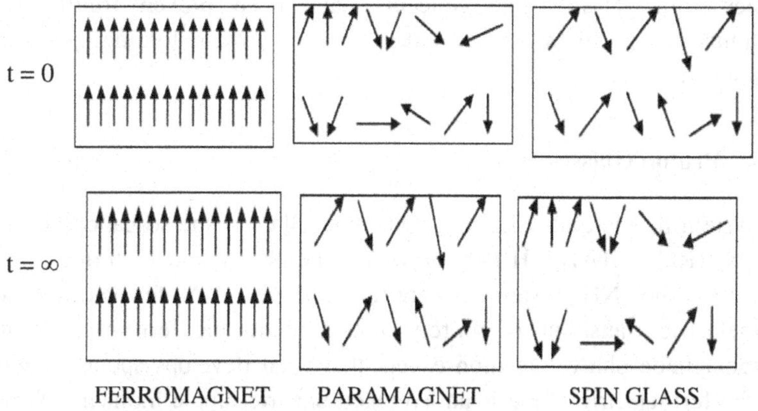

Fig. 3.3. A schematic picture of spin orientations at two different times: $t = 0$ and $t = \infty$.

What is the relation of q (t) to the usual Edwards–Anderson order parameter q? Note that the auto-correlation in Eq. (3.45) measures the probability that the ith spin points like $\vec{S_i}(t)$ at time t given that it points like $\vec{S_i}(0)$ at time t = 0. Thus, $< \vec{S_i}(0).\vec{S_i}(t) >$ measures spontaneous spin fluctuations for a system in overall thermal equilibrium. If we now, starting from the epoch t = 0, wait for an infinitely long time, we expect the memory (of the spin orientation) to be completely obliterated and therefore, the spin orientations at infinitely–separated times to be totally independent. This property of phase-space averaging is called 'Mixing' and it implies that [78]

$$\lim_{t\to\infty} < \vec{S_i}(0).\vec{S_i}(t) > = \ < \vec{S_i}(0) >.< \vec{S_i}(\infty) =< \vec{S_i} >^2. \qquad (3.46)$$

Applying this mixing property to Eq. (3.45) we conclude that

$$\lim_{t \to \infty} q\,(t) = q, \tag{3.47}$$

the Edwards–Anderson order parameter, introduced earlier (cf. (3.12)). The assumption of mixing under which the correlation function in Eq. (3.45) factors is obviously applicable with regard to thermal averaging and not to the full configuration averaging and therefore, the occurrence of a nonzero q in a glassy system is attributed to nonergodicity. Of course, even in ordinary ferromagnets, q is nonzero which, however, is the consequence of having a nonzero order in space, i.e., $< \vec{S_\iota} > \neq 0$, a feature that is absent in spin glasses. Other systems in which spin glass – like concepts have been proven fruitful are optimization problems in computer science, protein folding, polymer glasses, foams, etc.

3.5. Proton Glasses

A system that is quite akin to a Quantum Glass is the so-called Proton Glass, $(Rb)_{1-x}$ $(NH_4)_x$ H_2PO_4 in which the Rb-system is ferroelectric whereas the NH_4-system is antiferroelectric. A ferroelectric to paraelectric transition is entirely analogous to the ferromagnetic to paramagnetic phase transition except that what develops spontaneously in the ferroelectric phase is an averaged electric dipole moment. What causes this? Recall that the Rb–and the NH_4–systems are hydrogen-bonded materials in which the H can occupy two equivalent positions in the O-H ... O bonds [68]. The situation is similar to the one in which the H occupies one or the other of the two minima of a symmetric double-well potential. Now, when the H resides on one of the two minima it causes a local charge distortion thereby creating an electric dipole moment in a given direction. The dipole moment points in the opposite direction if the H resides on the other minimum. Thus the mean dipole moment, which is basically a scalar because of unidirectionality, is proportional to the average occupation of the H in one or the other minimum. If we map the occupation variable to an Ising spin that can take $+\frac{1}{2}$ or $-\frac{1}{2}$ values the mean dipole moment becomes proportional to

<S_z>. The latter would normally be zero because of the symmetry of the double-well but for a long ranged coupling between the distortion fields or the strain fields, because of the presence of the H at two distinct sites of the underlying lattice. Naturally this coupling will be of the Ising form but now between two "pseudo-spins", each pseudo-spin representing the occupation variable. From the point of view of the latter there will be a local mean field because of the Ising interaction which will cause an asymmetry in the double well, yielding a broken-symmetric ferro-or-anti-ferroelectric transition (cf., Chapter 1).

The nature of this transition will be ferroelectric, if the pseudo-spins prefer to be of the same sign or antiferroelectric if they alternate in sign from one H-site to another, just as in a magnetic case, which will again be determined by the negative or positive signs of the bond strengths, respectively. As in the case of the Magnetic Glass, the bond strengths J_{ij} are randomly distributed because of random admixture of Rb and NH_4 – systems, yielding both frustration and disorder effects. However, unlike the case of the Magnetic Glass, there is a local symmetry-breaking field in view of the fact that the replacement of Rb by NH_4 'weights' the two spin-states $+\frac{1}{2}$ and $-\frac{1}{2}$ differently. Consequently, the Hamiltonian may be written as

$$\mathcal{H} = -\sum_{i,j=1}^{N} J_{ij}\, S_{iz} S_{jz} - \sum_{i=1}^{N} f_i S_{iz}, \tag{3.48}$$

where, as stated above, the term f_i accounts for the local strain field due to substitution of, say NH_4 in place of Rb. Thus the underlying Hamiltonian is endowed by both random bonds and random fields.

The bond distribution is again governed by the probability given by Eq. (3.2) but now the mean value of J_{ij} is not zero but given by

$$J_0 = (1-x)(-J) + x(J) = (2x-1)J. \tag{3.49}$$

Similarly, the variance is

$$\Delta = (1-x)J^2 + xJ^2 - J^2(2x-1)^2 = 4x\,(1-x)J^2. \tag{3.50}$$

A similar (but independent) Gaussian distribution can be assumed for the random fields f_i [79].

Until now our discussion on the Hamiltonian in Eq. (3.48) is based entirely on classical considerations and the Hamiltonian is devoid of any quantum attribute, such as the transverse field term in Quantum Glasses. This indeed is the case when H is deuterated yielding a random admixture of $Rb_{1-x}(ND_4)_x D_2 HO_4$. The deuteron D, which is heavier than the proton H, can only traverse from one minimum of the double well to another if it is aided by thermal activation, which is a classical process. However, for the proton sample, dynamics can occur even at zero temperature, because the proton, being a light particle with relatively large de Broglie wavelength, can tunnel from one minimum to the other. Because the two states of the proton associated with the two minima are viewed as eigenstates of the spin operator S_z, tunneling causes admixture of the two eigenstates, and this can be effected by the operator S_x which is off-diagonal in the representation of S_z. The spin Hamiltonian for the proton glass can then be written by generalizing Eq. (3.48) as

$$\mathcal{H} = -\sum_{i,j=1}^{N} J_{ij} S_{iz} S_{jz} - \sum_{i=1}^{N} f_i S_{iz} - \Gamma \sum_{i=1}^{N} S_{ix}. \tag{3.51}$$

3.6. Structural Glasses

All the glassy systems we have discussed in sections 3.2 through 3.5 have one property in common: the disorder is a 'quenched' one because of the nature of preparation of the solid solution in which the impurity is randomly frozen-in. This is true in the case of the magnetic or the quantum glass in which yttrium is substituted in the holmium sites at random, as well as in the case of a spin glass in which the metallic copper is randomly replaced by magnetic manganese defects. On the other hand, for structural glasses like the ones found in an ordinary room-window, there is no externally imposed defect but merely 'dynamical quenching' of certain degrees of freedom because of the process of super-cooling of the underlying liquid. The system is then trapped in a metastable state from which nucleation to the stable state of a crystalline solid takes an infinitely long time. The result is a state of matter that has properties intermediate between those of liquids and crystalline solids. The liquid-like attribute is the complete absence of translational or orientational ordering of atoms whereas the solid-like property emerges from a non-vanishing shear modulus of elasticity.

In spite of such fundamental differences between the origins of structural glasses and spin glasses (and for that matter, magnetic glasses) the phenomenology of the two systems bears a remarkable similarity. For instance, one of the significant hallmarks of a structural glass, as mentioned in the preamble to this chapter, is the rapid rise in viscosity as one approaches the glass transition. For the so-called fragile liquids this increase is governed by the super-Arrhenius, Vogel-Fulcher law [58]:

$$\eta(T) \sim exp\left[\frac{A\,T_0}{T - T_0}\right], \tag{3.52}$$

where A is a constant. The temperature To is not experimentally reachable as it is about 10% lower than the glass transition temperature T_G at which the system freezes. When it comes to a spin glass there is no viscosity as such but there indeed is slow relaxation characterized by a relaxation time, which can be probed in a magnetization–decay or a frequency-dependent susceptibility measurement, that exhibits a Vogel-Fulcher relation. This analogy is quite appropriate if one recalls that the viscosity of a super-cooled liquid is proportional to the Maxwell time in a 'visco-elastic' sense [81].

Against this background of ultraslow relaxation being a common binding theme for structural glasses and spin glasses it is not surprising that theoretical developments in the two areas have also taken a common route. One of the phenomenological theories of a super cooled liquid which yields a fairly successful description of the growth of viscosity over a few decades of temperature is the so-called mode coupling theory [82]. This theory predicts the existence of what is referred to as an <u>ideal glass transition temperature</u> T_c at which the viscosity has a power law (and hence weaker than exponential) divergence. The temperature T_c is substantially larger than T_G but marks a crossover below which the super cooled liquid becomes extremely sluggish. A similar dynamical transition is exhibited by mean-field spin glass models with infinite ranged interactions discussed earlier which show non-ergodic behavior at a temperature T_d. Below T_d, at a temperature T_s ($T_s<T_d$), the configurational entropy per Ising spin goes to zero, and hence it is tempting to make a connection to glass transition by relating the dynamical spin glass transition at T_d to the ideal mode-coupling transition at T_c and the expected thermodynamic transition at T_s to the supposed entropy-vanishing transition for glasses, in the Kauzmann sense, occurring at T_0.

In addition to the identification of different threshold temperatures for structural glasses with those of spin glasses the analogy also extends into the domain of inhomogeneous local relaxation seen in super cooled liquids. This inhomogeneous relaxation which goes by the name of dynamical heterogeneity, can be captured in terms of a four-point correlation function that measures the spatio-temporal correlation between density relaxations at two spatially separated points [83]. Both numerical simulations for super cooled liquids and calculations for mean-field spin glass models display remarkably similar behavior of the dynamical heterogeneity.

With the preceding remarks it is evident that many of the ideas emerging out of the theories of spin glasses and magnetic glasses have been invoked in the much harder problem of structural glasses. Thus one frequently talks about the Edwards-Anderson order parameter q for structural glasses as well, though as has been mentioned earlier, there is no quenched disorder in super cooled liquids.

The underlying belief in this context is that it is the presence of frustration that is essential for glassy physics. Given this situation it is interesting to note that the replica formalism developed for the first time for magnetic spin glasses can be combined with standard liquid state theory to study the equilibrium properties of liquids under an external random potential [84]. The consequent analysis goes under the name of Replicated Liquid-State Theory. Just as in the replica theory of quenched disorder, $\alpha = 1, \dots . n$ replicas of the system are introduced and the $n \to 0$ limit is taken to obtain physical quantities averaged over different realizations of the disorder, in the replicated liquid-state theory one considers a mixture of n kinds of particle, each kind labeled by the replica index α. Averaging over the random potential yields a new attractive potential between particles of different species (or replicas).

Standard liquid state theories such as the hypernetted chain approximation is then generalized to treat this n-component liquid mixture and the $n \to 0$ limit is taken analytically. What then ensue are expressions for the equal time, two-point, disorder-averged local density correlation function as well as the disorder-induced, two-point but time-averaged local density, which compare favourably with simulations of real liquids in the presence of a random external potential [84]. A question of course naturally arises: what does a random potential have to do with super cooled liquids? This question is answered by Mézard and Parisi [85] who showed that the hypernetted chain approximation to a replicated liquid does admit one-step replica symmetry-breaking similar to the one found in mean-field spin glass models, discussed earlier. Thus, this one-step replica symmetry-breaking transition can be interpreted as a themodynamic transition for real glasses.

In conclusion however we should emphasize that although the glass transition has many common features with spin-glass-like models, the latter are all of the infinite-ranged mean-field types. These models therefore ignore the influence of fluctuations considered important for three-dimensional systems, especially with short-ranged forces.

Chapter 4

Relaxation Effects

4.1. Introductory Remarks

Our discussion on how developments in Magnetism and models for studying magnetic behavior have influenced phenomena in other fields has been restricted to effects in thermal equilibrium. Of equal, if not more topical interest, is in the investigation of time-dependent effects, the theory of which is based on Nonequilibrium Statistical Mechanics [86]. Here again Magnetism has played a key role in clarifying various issues that are germane to systems out of equilibrium, and these are the subject of the present and subsequent chapters.

Time-dependent effects come into reckoning in distinct physical contexts. The first and foremost actually concerns thermal equilibrium. For instance, consider again an Ising spin system in the paramagnetic phase and in equilibrium. The meaning of equilibrium is that average values of dynamical variables such as the spin magnetic moment are time-independent, in this case zero. However, if one tracks the time-development of an individual spin, the latter keeps on flipping from 'up' to 'down' states and vice-versa. The reason that the average magnetic moment, proportional to the magnetization, is zero in the paramagnetic phase is that the mean time for which the spin points up exactly equals the mean time for which the spin points down. It is precisely the non-zero difference between these two times which marks the onset of a paramagnetic to ferromagnetic transition. The most common experimental technique to probe these spontaneous (i.e. in equilibrium) spin flips is to subject the system to a weak external frequency-dependent magnetic field, wait for a sufficiently long time to ensure that the system has come to a steady state and measure spin correlations in time through a complex susceptibility [87]. Similar spontaneous fluctuations in time are of course also relevant for the velocity of a tagged particle in a Maxwellian gas in thermal equilibrium or position and velocity of a Brownian particle in a fluid in equilibrium or the jump diffusion of an interstitial defect via the vacancy mechanism in a solid, etc. It is evident therefore that studies of such spontaneous or intrinsic time-fluctuations of dynamical variables warrant nonequilibrium methods, albeit one is not considering a system out of equilibrium, per se. A magnetic paradigm of a single spin kinetics becomes handy in such considerations, as discussed in Sec. 4.2 below.

The next example considered in Sec. 4.3, relates to the response behavior of a system, initially in thermal equilibrium, to an external, static laboratory field. Once again, a convenient setting for elucidating the time-evolution of the system from the epoch $t = 0$ at which the external field is applied, is indeed the magnetic one. Here one measures the time-dependent magnetization response, which is now a nonlinear one in general (in contrast to the linear susceptibility response discussed above), and its asymptotic approach to a new thermal equilibrium. The time (or times) which characterize this approach is called relaxation time(s). A related experimental technique is to keep the field on for a long time such that equilibrium obtains and then switch the field off in order to investigate the time decay of say, the magnetization. What one learns in the magnetic context can be translated to similar phenomena in dielectrics, paraelectrics, anelastic relaxation due to defect motion in solids, etc [87].

The situation considered in the above paragraph refers to nonequilibrium effects occasioned by the application of an external field or removal of it in the laboratory. Such a laboratory perturbation can be effected by a magnetic field, an electric field, a stress field, a pressure, etc., as the case may be. On the other hand, studies of nonequilibrium phenomena are also important for systems far from equilibrium when certain internal fields are altered. Recall from Chapter 3 that it is precisely the process of super cooling or rapid lowering of the internal field such as the temperature that is eventually responsible for freezing into a glassy structure. Similar quenching techniques can be applied to a binary liquid mixture or an alloy which is suddenly brought down (in temperature) from a homogeneous phase to a mixed phase. The resultant phenomena go under the name of phase ordering kinetics or spinodal decomposition, a topic of great practical interest in Materials Science [20]. In Sec.4.4 below, we will show how the theory for the single spin kinetics developed in Sec. 4.2 can be generalized to an interacting spin case. e.g. for the Ising model and adapted for application to the distant topic of phase ordering [88]. In Sec. 4.5 we turn to quantum models such as the transverse Ising model, introduced earlier in Chapter 2, and discuss relaxation effects within mean field theory. Similar quantum spin relaxation theories, developed first in the context of electron spin resonance and nuclear magnetic resonance [89], find wide applications to laser spectroscopy and Quantum Optics [90]. The models in Magnetism, e.g. the transverse Ising model, further provides a platform for discussing competition between quantum effects and hysteresis, the latter being the hallmark of dissipative nonequilibrium phenomena [42]. Relaxation in glassy systems, in which one is confronted with the simultaneous presence of time-dependent effects and frustration combined with disorder, is yet another important phenomenon in which models developed for spin glasses find significant paradigmical applications. This is the subject of Sec. 4.6.

4.2. Single Spin Kinetics in Equilibrium

The physical system we want to treat is that of an isolated paramagnetic impurity which is assumed to have a spin equal to one-half. In equilibrium and in the absence of any kind of external field therefore, the spin magnetic moment denoted by S will keep flipping between two values + 1 or − 1 which are the projections of the moment along a laboratory fixed axis, say $\mathbf{z}$. Because the system is paramagnetic and there is no external bias the rate of transition from the state + 1 to − 1 must equal the rate of transition from − 1 to + 1, denoted by $\lambda/2$. Hence, if we consider P(s, t) which is the probability that the value of the spin moment is s (= ± 1) at a time t, its time-variation will have a 'loss term' for flipping out of the state s and a 'gain term' for flipping into the state s. Thus we may write down the simplest rate equation or master equation for P (s, t):

$$\frac{dP}{dt}(s,t) = -\frac{\lambda}{2}P(s,t) + \frac{\lambda}{2}P(\ \ s,t). \tag{4.1}$$

A similar rate equation applies of course to P(− s, t). Note that the corresponding equilibrium distributions, analogous to the continuum case of the Maxwellian probability distribution, are obtained by setting the left hand side of Eq. (4.1) to zero, thus

$$P_{eq}(s) = P_{eq}(-s). \tag{4.2}$$

Combining Eq. (4.2) with the conservation condition that

$$P_{eq}(s) + P_{eq}(-s) = 1, \tag{4.3}$$

yields

$$P_{eq}(s) = P_{eq}(-s) = \frac{1}{2}, \tag{4.4}$$

implying that the spin moment has as much probability of pointing up as pointing down, as would indeed be expected for a paramagnet.

We define that magnetization per spin as

$$m = \mu < S >, \tag{4.5}$$

where the parameter μ subsumes the Bohr magneton and the g-factor and the angular brackets $< \cdots >$ denote average over the probability distribution of Eq. (4.1). Clearly, the rate of change of m in this simple model would be given by

$$\frac{dm(t)}{dt} = \mu\frac{d}{dt} < S > = \mu\frac{d}{dt}\sum_{s=\pm 1} sP(s,t) = \mu\sum_{s=\pm 1} s\frac{d}{dt}P(s,t). \tag{4.6}$$

Substituting Eq. (4.1) and switching the sign of s in the second term we can easily derive

$$\frac{dm(t)}{dt} = -\lambda\, m(t). \tag{4.7}$$

This is how the magnetization, i.e. the time-averaged spin magnetic moment of a tagged atom, fluctuates in time, spontaneously. On the other hand, in equilibrium,

$$m_{eq} = \mu < s >_{eq} = \mu \sum_{s=\pm 1} sP_{eq}(s) = 0, \tag{4.8}$$

from Eq. (4.4). Therefore, Eq. (4.7) implies that, starting from an arbitrary initial condition: $m(t = 0)$, the magnetization relaxes to the equilibrium value of zero, in accordance with

$$m(t) = m(t = 0)e^{-\lambda t}. \tag{4.9}$$

Equation (4.9) allows us to define an important physical attribute of the system: the relaxation time τ which for the present model is given by

$$\tau = \frac{1}{\lambda}. \tag{4.10}$$

The strategy of making an equilibrium measurement in the laboratory is to wait for a sufficiently long time t_{exp} such that $t_{exp} \gg \tau$, where the subscript 'exp' stands for experiment. One other way of looking at the issue is to time-average m(t), denoted by an overhead bar, over the 'time-window' τ_{exp} of the measurement, thus yielding

$$\bar{m}(t) = \frac{1}{\tau_{exp}} \int_0^{\tau_{exp}} m(t)dt = m(o)\frac{\tau}{\tau_{exp}}(1 - e^{-\tau_{exp}/\tau}), \tag{4.11}$$

which again vanishes for ${}^{\tau_{exp}}/_{\tau} \gg 1$. The fact that the time average is the same as the ensemble average (cf. Eq.(4.8)) is an expression of a property, called Ergodicity. There must be certain coupling terms between the spin of interest and other observables of the heat bath. The most common such observables for a magnetic solid are the phonon fields which describe thermal vibrations of the lattice. The above simple model then epitomizes what is called spin-lattice relaxation [91]. What these phonon fields are will be discussed in somewhat more detail in Chapter 6 later.

We may also remark that the mathematical premise under which the rate equation (4.1) is written down is the simplest example of a discrete stochastic process that goes under the name of a two-state jump process or a telegraph process [87]. Thus a time-sampling of the stochastic variable s looks like a telegraphic signal as sketched in Fig. 4.1.

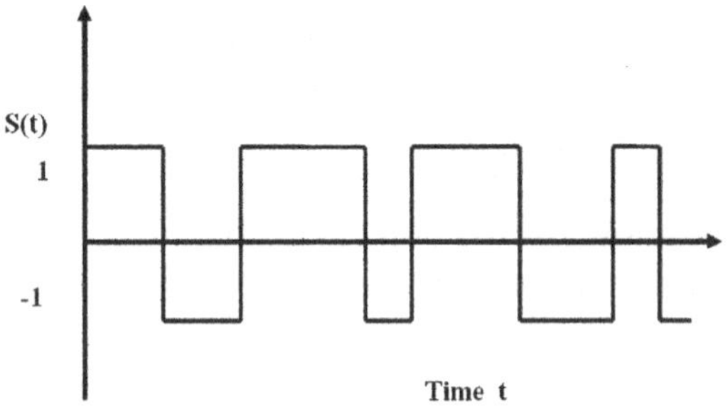

Fig. 4.1. A random sampling of the stochastic process s(t).

At random instants of time s(t) dichotomously jumps from + 1 to – 1 and vice-versa. However, the mean time for which s(t) has the value +1, obtained by averaging over all the horizontal segments above the abscissa is the same as the mean time for which s(t) has the value – 1, obtained by averaging all the horizontal segments below the abscissa.

How does one probe these spontaneous spin fluctuations in the laboratory? As mentioned in the introductory remarks of Sec. 4.1, one applies a weak monochromatically oscillatory field of frequency ω, along the z-axis and observes the liner response to that field encapsulated by the complex susceptibility $\chi(w)$ given by [87] :

$$\chi(w) = -i\omega\mu^2 N\beta \lim_{\delta\to 0} \int_0^\infty dt e^{-\delta t + i\omega t} \left[< S_z^2 >_{eq} - < S_z(o) S_z(t) >_{eq}\right], \tag{4.12}$$

where $\beta = (k_B T)^{-1}$, and N is the number of spins in the sample. The quantity $< S_z(o)S_z(t) >_{eq}$ or simply $< s(o)s(t) >_{eq}$ *in Eq.*(4.12) measures the correlation between the z-component of the spin at two different times 0 and t, and is appropriately referred to as the spin-auto correlation function. On the other hand $< S_z^2 >_{eq}$ may be interpreted as equal-time auto correlation function which, in the present case, is simply unity because we are dealing with a spin one-half entity. Thus

$$\chi(w) = -i\omega N\mu^2\beta \lim_{\delta\to 0} \int_0^\infty dt e^{-\delta t + i\omega t} \left(1 - < s(o)s(t) >_{eq}\right). \quad (4.13)$$

How does one calculate $< s(o)s(t) >_{eq}$? Well, what P(s, t) measures is the single-point probability function for an underlying process that is called a Stationary Markov Process, stationarity is implied by the fact that the relevant correlation function depends only on the difference between the two times that occur in the argument of the variable s(t). The other function one needs is a two-point function which is the conditional probability $P\,(s'|s;t)$ which measures the probability that the spin takes the value s at time t given that it has the value s' at time $t = 0$. It is evident that the conditional probability obeys an identical rate equation as (4.1):

$$\frac{d}{dt} P(s'|\, s\,; t) = \frac{-\lambda}{2} P(s'|s\,; t) + \frac{\lambda}{2} P(s'| - s\,; t). \quad (4.14)$$

It is easy to then write down the correlation function in equilibrium:

$$< s(o)s(t) >_{eq} = \sum_{s',\ s} P_{eq}(s')\, s' P(s'|\, s\,; t)\, s. \quad (4.15)$$

The physical meaning of Eq. (4.15) is that starting from a random value s' picked out of the equilibrium ensemble formed by the probability function $P_{eq}(s')$, $P(s'|\, s\,; t)$ measures the propagation in time t to a new value s, and both s' and s must be summed over all possibilities. Correspondingly, from Eqs. (4.14) and (4.15),

$$^d/_{dt} < s(o)s(t) >_{eq}$$

$$= -{}^{\lambda}/_2 \sum_{s,\, s'} \{P_{eq}(s')s's\, P(s'|s; t) - P_{eq}(s')s's\, P(s'| - s; t)\}$$

$$= -\lambda < s(o)s(t) >_{eq}, \quad (4.16)$$

where in the second term on the right hand side we have again flipped the sign of s, as before. Not surprisingly, the correlation function in equilibrium follows the same equation as the magnetization (cf., Eq. (4.7)).

We may also point out an important property that the correlation function must satisfy as the argument t goes to infinity. Note from the defining equation (4.15) that

$$\lim_{t\to\infty} < s(o)s(t) >_{eq} = \sum_{s',s} P_{eq}(s')s's \lim_{t\to\infty} P(s'|s\,;t). \qquad (4.17)$$

From physical grounds it is expected that

$$\lim_{t\to\infty} < s(o)s(t) >_{eq} = \left(\sum_{s'} \mathrm{P}_{\mathrm{eq}}^{(s')s'}\right)\left(\sum_{s} P_{eq}(s)s\right) = < s(o) >_{eq}^{2}. \qquad (4.18)$$

Although the latter quantity on the right of Eq. (4.18) is zero in the present case, the property that the correlation function factors in the long-time limit into product of averages is a general one that goes under the name of "Mixing". Mixing is a stronger condition than ergodicity and in fact subsumes the latter [78]. With these inputs the solution of Eq. (4.16) reads

$$< s(o)\, s(t) >_{eq} = e^{-\lambda t}, \qquad (4.20)$$

where again, we have used the spin-half property ($i.e.$ $s = \pm 1$) to ensure that the initial value is unity.

With Eq. (4.20) at hand the susceptibility in linear response theory can be worked out from Eq. (4.13) as

$$\chi(\omega) = N\beta\mu^2 \frac{1}{1 - i\omega\tau}\,. \qquad (4.21)$$

Clearly, the susceptibility has a real part:

$$\chi'(\omega) = \frac{N\beta\mu^2}{1 + \omega^2\tau^2}, \qquad (4.22)$$

and an imaginary part:

$$\chi''(\omega) = N\beta\mu^2 \frac{\omega\tau}{1+\omega^2\tau^2}, \tag{4.23}$$

the two being related by the Kramers Krönig formula [89].

Though Eq. (4.23) is derived for the special model of a single spin relaxation the form of the ω-dependent susceptibility is as generic as the Drude conductivity for a metal [92], and is applicable to a variety of other contexts e.g. dielectric relaxation [93], strain relaxation due to diffusion of point defects [94], etc. A few general conclusions can then be drawn:

The relaxation is characterized by just one parameter τ that is variously called the relaxation time, collision time, scattering time, etc. Appropriately the response goes under the name of Debye response, first talked about in the context of dielectric relaxation, that goes hand in hand with the exponential decay in time (cf., Eq. (4.20). Any deviation from Debye relaxation points to complexity as seen for instance in glassy systems.

The time-window, which was earlier discussed in terms of the experimental time t_{exp}, is now provided by ω^{-1}. When that window is very long i.e. ω is vanishingly small, the system has time to equilibrate and hence $\chi(\omega)$ reduces to the equilibrium susceptibility:

$$\chi\,(\omega = 0) = \chi_0 = N\mu^2, \tag{4.24}$$

The entity $\chi''(\omega)$ governs dissipation in the system. Plotted as a function of ω it exhibits a symmetric peak called the Debye peak at $\omega = \tau^{-1}$, which is a Lorentzian with a width given by

$$\Gamma = 2\sqrt{3/\tau}. \tag{4.25}$$

Therefore, a measurement of the width yields the dependence of τ on other physical parameters, e.g. temperature, pressure, etc. For instance if relaxation is governed by a thermally activated process, τ would be expected to be given by the Arrhenius formula:

$$\tau = \tau_0 \exp\left(U/k_B T\right), \tag{4.26}$$

where U is the barrier energy that has to be overcome.

The assumed temperature-dependence of τ as in Eq. (4.26) allows us to obtain another glimpse into the time-window effect. Normalizing $\chi'(\omega)$ by χ_0 (cf. Eqs. (4.22) and (4.24)) and plotting the resultant quantity versus the temperature T (cf. Fig. 4.2) we obtain two different curves for two distinct ω's [95]. Both the curves exhibit Curie law ($\chi \propto \frac{1}{T}$) at high temperatures, but for $\omega = 100\ Hz$ (the higher of the two ω-values considered), the time-window is relatively smaller and therefore, below a certain temperature, the relaxation is comparatively shower $\left(i.e.\,\tau > \frac{1}{\omega}\right)$. As a consequence, there is deviation from the Curie law – the susceptibility actually starts dropping below a temperature called the 'Blocking' temperature T_B, which will be discussed in more detail in the next chapter. Below T_B the spin is not allowed to relax within the prescribed time-window, i.e. the spin is blocked. Naturally, for the smaller frequency $w = 10\ Hz$, when the time-window is relatively larger, T_B shifts to lower temperature as one needs to slow down the relaxation further.

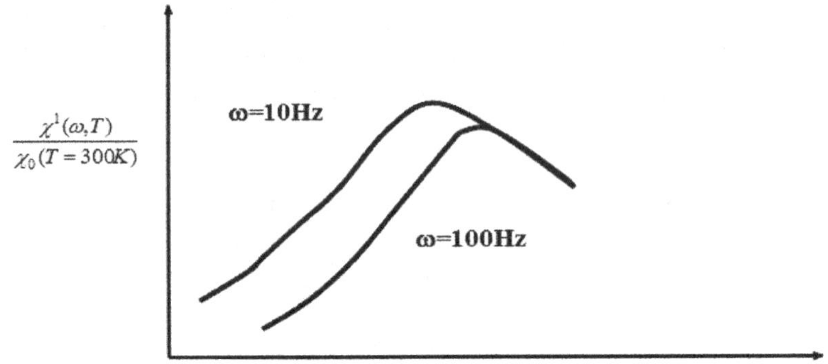

Fig. 4.2. $\chi'(\omega)$ vs T with $\frac{u}{k_B} = 8.00$K and $\tau_0 = 10^{-10}$ sec.

4.3. Non-Equilibrium Response of a Single Spin

In this section we consider the same spin-half paramagnetic impurity, initially in thermal equilibrium with the surrounding heat bath, but has its equilibrium disturbed by the application of a static magnetic field H applied along the **z**-axis at time $t = 0$. Clearly, if the field is kept on for a sufficiently long time, the impurity spin is expected to come to a new thermal equilibrium, at the same temperature of the bath but now governed by different probability distributions. Thus, at time $t = 0$,

$$P_{eq}(\pm s) = \frac{1}{2}, \tag{4.27}$$

as in Eq. (4.4), but at time $t = \infty$,

$$P_{eq}(\pm s) = \frac{e^{\mp\beta\mu Hs}}{\sum_{s=\pm1} \bar{e}^{\beta\mu Hs}} = \frac{e^{\mp\beta\mu Hs}}{2\cosh(\beta\mu H)}. \tag{4.28}$$

Our purpose in this section is to analyze how the system develops from the conditions described by Eq. (4.27) to the conditions described by Eq. (4.28). Albeit the discussion is carried out within the paradigm of a magnetic system very similar considerations would apply to a paraelectric subject to an external electric field or a paraelastic subject to an external stress.

The time evolution of the probability is now dictated by the rate equation (cf., Eq. (4.1)):

$$\frac{d}{dt}P(s,t) = -\lambda_{s,-s}\, P(s,t) + \lambda_{-s,\ s}\, P(-s,t), \tag{4.29}$$

where the rates of transition satisfy the so-called detailed balance condition:

$$\lambda_{s,-s}\, P_{eq}(+s) = \lambda_{-s,\ s}\, P_{eq}(-s), \tag{4.30}$$

where $P_{eq}(\pm s)$ are now given by Eq. (4.28) and not (4.27)! Thus the ratio of the two rates can be written from Eq. (4.28) by

$$\frac{\lambda_{s,-s}}{\lambda_{-s,\ s}} = e^{2\beta\mu Hs}. \tag{4.31}$$

We may choose

$$\lambda_{-s,s} = \lambda P_{eq}^{(+s)}, \tag{4.32}$$

where λ is assumed to be independent of s such that Eq. (4.32) reduces to the rate constant of the zero-field case discussed in Sec. 4.2.

We are now in a position to evaluate the time-development of the magnetization, induced by the external magnetic field. Note that (cf., Eq. (4.6)).

$$\frac{d}{dt}m(t) = \mu \sum_{s=\pm 1} s\left[\lambda_{s,-s}P(-s,t) - \lambda_{-s,\ s}P(s,t)\right]$$

$$= \lambda\,\mu \sum_{s=\pm 1} s\left[P_{eq}(s)P\,(-s,t) - P_{eq}(-s)P\,(s,t)\right]$$

$$= \lambda\,\mu \sum_{s=\pm 1} s\,P_{eq}(s) - \lambda\,\mu \sum_{s=\pm 1} s\,P\,(s,t), \tag{4.33}$$

where in the last line we have made use of the fact that conservation of probability works at all times, as it must, i.e.,

$$P_{eq}(s) + P_{eq}(-s) = 1,$$

and

$$P(s,t) + P\,(-s,t) = 1. \tag{4.34}$$

Therefore,

$$\frac{d}{dt}m(t) = -\lambda\left[m(t) - m_{eq}\right], \tag{4.35}$$

where, in the present case,

$$m_{eq} = \mu \sum_{s} sP_{eq}(s) = \mu \tanh(\beta\mu H). \tag{4.36}$$

The bulk magnetization M(t) then (defined as M(t) = Nm(t)) evolves in accordance with

$$\frac{dM(t)}{dt} = -\lambda[M(t) - N\mu \tanh(\beta\mu H)], \tag{4.37}$$

where N is the number of spins in the sample. The solution of Eq. (4.37), appropriate to the boundary conditions, reads

$$M(t) = N\mu\ \tanh(\beta\mu H)(1 - e^{-\lambda t}). \tag{4.38}$$

It is evident that the magnetization M(t) asymptotically approaches a value appropriate to the new equilibrium, and given by

$$M(\infty) = N\mu \tanh(\beta\mu H). \tag{4.39}$$

Interestingly, the same relaxation time $\tau\,(=\lambda^{-1})$ governs the approach to the new equilibrium as it does for spontaneous fluctuations, discussed in Sec. 4.2. This has to do with our assumption, stated below Eq. (4.32), that λ is independent of the state of the system. This assumption is not generally true, as will be discussed in the next chapter.

We want to first discuss the low-H behavior, i.e. linear response. In that limit Eq. (4.38) reduces to

$$M(t) = \beta N\,\mu^2 H\left(1 - e^{-t/\tau}\right), \tag{4.40}$$

wherein λ has to be independent of H too, and is written as the inverse of the relaxation time τ (cf. Eq. (4.10). In this regime we can define a linear, time-dependent susceptibility or a response function, given by

$$\chi(t) = \frac{M(t)}{H} = \beta N \mu^2 \left(1 - e^{-t/\tau}\right). \tag{4.41}$$

Interestingly, the steady-state frequency dependent susceptibility derived in the previous section (cf., Eq. (4.21) can be obtained from $\chi(t)$ by taking its Laplace transform, with the aid of the formula:

$$\chi(\omega) = i\omega \lim_{\delta\to o} \int_o^\infty dt e^{-\delta t + i\omega t}\,\chi(t). \tag{4.42}$$

Equation (4.42) which embodies the relaxation-response relation that connects the fluctuations in equilibrium with the nonequilibrium response to a weak external field is an essential consequence of the linear response theory [87].

Coming back to the nonlinear response captured by Eq. (4. 38), we present a sketch in Fig. 4.3 in which M(t) is plotted versus time t for two different values of the temperature T, after assuming a T-dependent τ as in Eq. (4.26). Quite interestingly, while the asymptotic response in equilibrium given by Eq. (4.39) is larger at the lower temperature T_1, approaching the saturation value of $N\mu$ at T = 0, and smallest at higher temperatures in accordance with the Curie law, the initial rise is steeper for the higher temperature T_2 because of the exponential dependence of τ on T as given by Eq. (4.26). Thus the different curves cross implying that there is a temperature regime in which the magnetization actually increases with the increase of the temperature! This counter-intuitive result holds only when the system is caught between two equilibrium status (at $t = 0$ and $t = \infty$) and has an important consequence for memory and aging effects seen in glassy systems (Chapter 5). Thus this simple two-state model is able to capture an essential ingredient of a much more complex behavior occasioned by a distribution of relaxation times τ due to interaction effects, frustration and disorder.

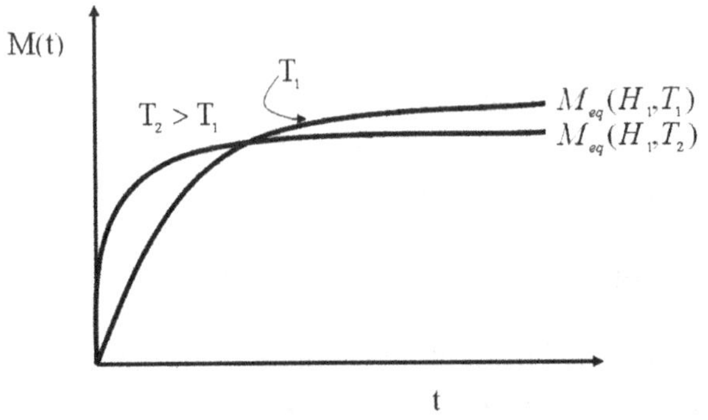

Fig. 4.3. Magnetization response vs. time t at two different temperatures.

Before closing this section, we consider yet another kind of experiment in which an external field H is kept on for a sufficiently long time such that the magnetization reaches its asymptotic value and then the field is switched off. The resultant time-decay, called the Magnetic After Effect, is again discussed most conveniently within the linear response theory, and is given from Eq. (4.40) by

$$M(t) = M_o \exp(-t/\tau), \tag{4.43}$$

where

$$M_o = \beta N \mu^2 H. \tag{4.44}$$

Once again the approach to equilibrium, in which the magnetization relaxes to zero, is expected to occur at different rates, at different temperatures. The counterpart of magnetic after effect is called the elastic after effect in the context of anelastic relaxation [94].

4.4. Effects of Interaction on Relaxation

We have mentioned in the earlier chapters that the Ising model, though introduced in the context of Magnetism, finds wider applications to Materials Science in relation to phase separation and phase ordering phenomena seen in binary liquid mixtures, alloys, etc [88]. We now discuss relaxation behavior of the Ising model which has direct relevance to these problems and which can be viewed as a natural generalization of the single spin kinetics discussed above. Thus we imagine our Ising spin (or pseudo spin) system to be in contact with an (invisible) heat bath which drives thermal fluctuations into the spin system causing the spins to flip at random instants of time.

When it comes to physical processes however, two distinct kinds of relaxation mechanisms need to be discussed. If it is the kinetics between the paramagnetic to the ferromagnetic phases of a magnet that is under study, relaxation can be envisaged to proceed via single spin flips. That will surely not conserve the total magnetization and hence the accompanying process is said to yield Nonconserved Kinetics.

On the other hand, if the Ising model is viewed to represent an AB-binary alloy, kinetics must proceed via diffusive jumps that exchange A and B atoms in their respective sites – a process that is usually mediated by a vacancy, assumed present in low concentrations. In addition, if like atoms (A or B) have more affinity for each other than for unlike atoms, i.e. $\varepsilon^{AA}(\varepsilon^{BB}) > \varepsilon^{AB}$, where epsilon ($\varepsilon$) represents energy of attraction, the net Ising exchange energy is 'ferromagnetic' ($i.e.\ J_{ij} > 0$) in nature. Thus the ordered phase is one of phase separation in which one finds AA and BB – clusters. In this situation, the relevant order parameter which is the difference between the number of A and B atoms, is conserved, because there is no net loss of atoms in the diffusion process. In the spin language relaxation occurs by a 'flip-flop' mechanism in which if a given spin is flipped down from the 'up' state a neighboring spin must be 'flopped' up from the 'down' state, consequently giving rise to

Conserved Kinetics. It is interesting to note that although it is the same Ising model that underlines the physical interaction terms for both the paramagnetic to the ferromagnetic transition in magnets as well as uniform to phase separated states in alloys (and hence characterizes an identical equilibrium behavior) the relaxation or approach to a new equilibrium proceeds via very different routes. Thus, for instance, how the domain-sizes grow in time during the coarsening processes is quite different for nonconserved and conserved kinetics, as discussed below.

a. Nonconserved Kinetics

At any given point in time the spin configuration is specified by a quantity: $\{s\} = \{s_1, s_2, s_3, \dots\dots\dots . s_N\}$ which registers that the spin at site 1 is in the state s_1, (where s_1 can be either +1 or - 1), spin at site 2 is in the state $s_2, \dots\dots\dots\dots$ and the spin at the Nth site is in the state s_N. The probability $P(\{s_i\}, t)$ that this configuration occurs satisfies a master equation which is a straightforward generalization of Eq. (4.29)

$$\frac{d}{dt}P(\{s\},t) = -\sum_{j=1}^{N} W\left(s_1, s_2, \dots\ s_j \dots s_N \middle| s_1, s_2, \dots -s_j, \dots . s_N\right)P(\{s,\},t)$$

$$+\sum_{j=1}^{N} W\left(s_1, s_2, \dots -s_j, \dots s_N \middle| s_1, s_2, \dots\ s_j, \dots . s_N\right)P(\{s,\}',t), \qquad (4.45)$$

where the configuration $\{s\}'$ denotes the state $\{s\}$ with $s_j \to -s_j$. As earlier, the underlying stochastic process is assumed to be Markovian and stationary (because the transition probabilities W are taken to be time-independent). Equation (4.45) goes under the name of the Glauber model [96]. The transition probabilities W will have to of course satisfy the detailed balance condition as in Eq. (4.30). Recall that the latter, from Eqs. (4.32) and 4.28), can be recast as

$$\lambda_{-s,+s} = \frac{\lambda}{2}[1 - \tanh(\beta\mu HS)], \qquad (4.46)$$

which, upon using the fact that $s = \pm 1$, can be reexpressed as

$$\lambda_{-s,+s} = \frac{\lambda}{2}[1 - s\ \tanh(\beta\mu HS)]. \tag{4.47}$$

In the present case of an interacting system the local magnetic field, say at the ith site and denoted by H_i, can be written as

$$\mu H_i = \mu h_i + \sum_{j=1}^{N} J_{ij} S_j, \tag{4.48}$$

where, for a binary AB – alloy,

$$h_i = \frac{1}{4} \sum_{j=1}^{N} (\varepsilon_{ij}^{AA} - \varepsilon_{ij}^{BB}), \tag{4.49}$$

and

$$J_{ij} = \frac{1}{4} (\varepsilon_{ij}^{AA} + \varepsilon_{ij}^{BB} - 2\varepsilon_{ij}^{AB}). \tag{4.50}$$

Generalizing Eq. (4.47) then, we may write

$$W(s_1, s_2, \dots -s_j, \dots s_N | s_1, s_2, \dots s_j, \dots s_N) = \frac{\lambda}{2}[1 - s_j \tanh(\beta\mu H_j)]. \tag{4.51}$$

Correspondingly, from Eqs. (4.35) and (4.36) we may read the equation of motion for the averaged spin as

$$\lambda^{-1} \frac{d}{dt} < s_i(t) >= -< s_i(t) > +< \tanh \beta(\mu h_i + \sum_{j=1}^{N} J_{ij} s_j) >. \tag{4.52}$$

The above equation is however analytically intractable except in the special case of one dimension because the second term on the right results in a hierarchical set of correlation functions, as can be seen by expanding the tanh-function [96]. The hierarchy can be truncated by invoking the mean field approximation, discussed earlier in chapters 2 and 3. The latter assumes that there is no correlation between different sites, i.e. the average of the product of spin operators can be replaced by the product of their averages. The result of such a random-phase decoupling is that the angular brackets denoting the statistical average can be taken inside the argument of the tanh-function [97]. Thus we find for the order parameter,

$$\lambda^{-1}\frac{d}{dt}m_i = -m_i + \mu \tanh[\beta(\mu h_i + \sum_{j=1}^{N} \widetilde{J_{ij}}\, m_j)], \qquad (4.53)$$

where the tilde on $\widetilde{J_{ij}}$ implies a division by μ. Even after making the mean-field approximation, Eq. (4.53) is highly nonlinear. Further simplifications ensue upon coarse-graining of the equations, in which the discrete nature of the lattice on which the spins are located is ignored by making the reasonable assumption that the order parameter is a slowly-varying (continuous) function of space and time [88].

For simplicity we consider the case of zero magnetic field which is tantamount to assuming that $\varepsilon^{AA} = \varepsilon^{BB}$ (cf. Eq. (4.49). In the continuum limit, the interaction range R can be defined as

$$R^2 = [J(o)]^{-1}\sum_{j=1}^{N} |(\vec{r_i} - \vec{r_i})|^2 \, J_{ij}, \qquad (4.54)$$

where $\vec{r_i}$ in the position vector of site i, and

$$J(o) = \sum_{j=1}^{N} \tilde{J}_{ij}. \qquad (4.55)$$

We can now expand terms as follows:

$$\sum_{j=1}^{N} \tilde{J}_{ij}\, m_j \simeq J(o)[m(\vec{r_i}, t) + \frac{1}{2}R^2\, \nabla_i^2 m(\vec{r_i}, t|] + . \qquad (4.56)$$

and

$$\tanh[\beta\sum_{j=1}^{N} \tilde{J}_{ij}\, m_j] \simeq \frac{T_c}{T}m(\vec{r_i}, t) - \frac{1}{3}(\frac{T_c}{T})^3 m(\vec{r_i}, t)^3$$

$$+\frac{1}{2}\frac{T_c}{T}\, R^2\, \nabla_i^2\, m(\vec{r_i}, t) + \ldots\ldots\ldots, \qquad (4.57)$$

where

$$T_c \equiv \frac{J(o)}{K_B}. \tag{4.58}$$

Equation (4.53) then yields, after dropping the subscript i for the position variable,

$$\lambda^{-1}\frac{\partial}{\partial t}m\,(\vec{r},t) = \left(\frac{T_c}{T}-1\right)m(\vec{r},t) - \frac{1}{3}\left(\frac{T_c}{T}\right)^3 m(\vec{r},t)^3$$

$$+\frac{T_c}{2T}\,R^2\,\nabla^2\,m\,(\vec{r},t) + \text{ other terms} \tag{4.59}$$

which is in the form of a time-dependent Ginzburg – Landau (TDGL) equation:

$$\lambda^{-1}\frac{\partial}{\partial t}m\,(\vec{r},t) = -\frac{\delta G[m]}{\delta m}, \tag{4.60}$$

where ${}^{\delta G}/_{\delta m}$ denotes the functional derivative of the free-energy functional:

$$G[\mathrm{m}] = \int d\vec{r}\,[G(m) + \frac{1}{2}K(\vec{\nabla}m)^2], \tag{4.61}$$

appropriate to the <u>local</u> free energy:

$$\beta G(m) = -\frac{1}{2}\left(\frac{T_c}{T}-1\right)m^2 + \frac{1}{12}\left(\frac{T_c}{T}\right)^3 m^4, \tag{4.62}$$

and the ‘surface tension’ parameter:

$$K = \mathrm{k}_B T_c\,R^2. \tag{4.63}$$

Although the justification of neglecting the higher order terms in the above derivation is valid only for $T \rightarrow T_c$ when the order parameter is small, it is generally believed that the TDGL equation (4.60) is valid even for deep quenches. Therefore, notwithstanding the magnetic context of the Ising model, the TDGL equation is universally accepted as a reasonable model for a large class of nonconserved ordering processes. Consider then a rapid quench from the time $t = o$ when the system is brought from a disordered state ($T > T_c$) to a state below T_c. Because the disordered state is no longer the preferred equilibrium state at $T < T_c$, the far-from-equilibrium homogeneous system evolves towards its new equilibrium state by separating into equivalent domains for positive and negative values of the order parameter (see Fig. 4.4). These domains coarsen with time and are characterized by a growing length scale $L(t)$. A finite system becomes ordered in either of two equivalent states as $t \to \infty$. In dimensions greater than one the domain length scale obeys the Lifshitz–Allen–Cahn (LAC) law, $L(t) \sim t^{1/2}$ which can be understood as follows. Denoting the order parameter values as $m = \pm 1$ for the bulk domains and as $m = o$ for the interfaces, the interfacial velocity υ in the direction of increasing m is determined by $|\upsilon| \sim \varkappa$, where $\varkappa$ is the local curvature. Now, $\upsilon = \frac{dL}{dt}$ whereas $K \sim \frac{1}{L}$ and therefore, $L(t) \sim t^{1/2}$, the LAC law.

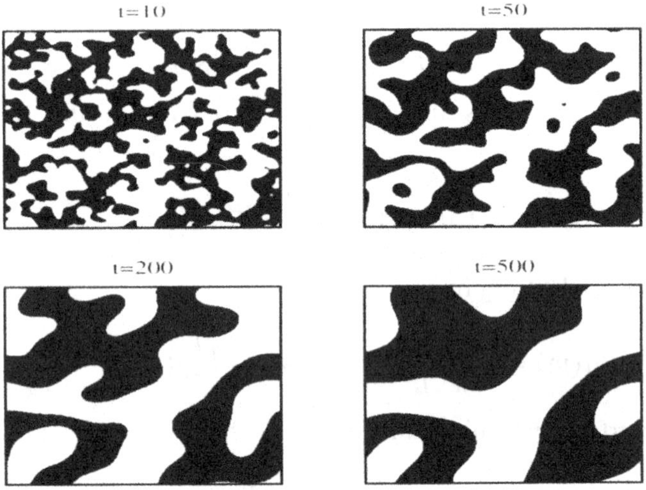

Fig. 4.4. Simulation of Eq. (4.60). Black regions denote positive magnetization while empty regions describe negative magnetization (after Ref. [88]).

b. Conserved Kinetics

The single-spin Glauber model as described above is appropriate for nonconserved kinetics seen for instance in magnets. But, as discussed earlier, when it comes to a binary AB – alloy, either exhibiting phase separation ($J_{ij} > o$) or order disorder transitions in which A and B atoms order in alternate lattice sites ($J_{ij} < o$), the net number of atoms remains constant. Hence the order parameter defined by $(n_A - n_B)$, which is the difference between the two net concentrations, is conserved. We are thus led to the Kawasaki model [98] which envisages a spin flip-flop process, alluded to earlier.

The corresponding master equation can be written as

$$\frac{d}{dt}P(\{s_i\},t)$$

$$= -\sum_{j=1}^{N}\sum_{k\epsilon L_j} W\left(s_1, \ldots . s_j, s_k, \ldots . s_N \,\middle|\, s_1, \ldots s_k, s_j, \ldots . s_N\right) P(\{s_i,\},t)$$

$$+\sum_{j=1}^{N}\sum_{k\epsilon L_j} W\left(s_1, \ldots . s_k, s_j, \ldots . s_N \,\middle|\, s_1, \ldots s_j, s_k, \ldots . s_N\right) P(\{s_i,\}',t). \quad (4.64)$$

In this case, the stochastic process involves, e.g., exchange of s_j (or s_k) at site j with s_k (or s_j) at site k $\in L_j$, where L_j denotes a neighboring site of j. In the second term on the right hand side of Eq. (4.64), $\{s_i\}'$ refers to the configuration obtained from $\{s_i\}$ by interchanging $s_j \leftrightarrow s_k$. We may note in passing that while the Glauber model connotes to spin-lattice relaxation the Kawasaki model is based on a spin-exchange process that can be induced by a dipolar coupling between spins. The process is well known in Magnetism in the context of 'exchange narrowing'[99] and 'motional narrowing' phenomena, seen in magnetic resonance [100].

Assuming again that the magnetic field is zero ($\varepsilon^{AA} = \varepsilon^{BB}$, for the alloy problem) the detailed balance condition for the transition probabilities can be expressed as

$$W\left(s_1, \ldots . s_j, s_k, \ldots . s_N \,\middle|\, s_1, \ldots s_k, s_j, \ldots . s_N\right)$$

$$= \frac{\lambda}{2}\{1 - \tanh[\frac{\beta}{2}(s_j - s_k)(\sum_{n\epsilon L_j'} J_{jn}\, s_n - \sum_{n\epsilon L_k'} J_{kn}\, s_n)]\}. \quad (4.65)$$

The primes in the summations on the right hand side of Eq. (4.65) denote constraints as $n \neq k$ (for n $\epsilon L'_j$) and $n \neq j$ (for $n\epsilon L_k'$).

Focusing on the nearest neighbor coupling only in the underlying Hamiltonian, we note that $(s_j - s_k) = 0, \pm 2$ and hence, we can extract this factor from the argument of the tanh-function to obtain

$$W(s_1, \ldots . s_j, s_k, \ldots . s_N \,|s_1, \ldots s_k, s_j, \ldots . s_N)$$

$$= \frac{\lambda}{2}\{1 - \frac{s_j - s_k}{2} \tanh[\beta \sum_{n \epsilon L'_j} J_{jn}\, s_n - \sum_{n \epsilon L'_k} J_{kn}\, s_n]\}. \quad (4.66)$$

As before we concentrate on the time evolution equation for the order parameter which is obtained by multiplying both sides of Eq. (4.64) with s_i and summing over all possible configurations, yielding

$$2\lambda^{-1} \frac{d}{dt} < s_i(t) > = -\xi < s_i > + \sum_{k \varepsilon L_i} < s_k >$$

$$+ \sum_{k \varepsilon L_i} < (1 - s_i\, s_k) \tanh \left[\beta(\sum_{n \epsilon L'_i} J_{in}\, s_n - \sum_{n \epsilon L'_k} J_{kn}\, s_n) \right] >, \quad (4.67)$$

where ξ is the number of nearest – neighbours (co-ordination number) of a lattice site. We again invoke the mean field approximation that amounts to replacing the expectation value of the function on the right hand side of Eq. (4.67) by the function of the expectation values. The result is

$$2\lambda^{-1} \frac{d}{dt} < s_i(t) > = -\xi < s_i > + \sum_{k \epsilon L'_i} < s_k >$$

$$+ \sum_{k \varepsilon L_i} (1 - < s_i > < s_k >)$$

$$\times \tanh \left[\beta(\sum_{n \epsilon L'_i} J_{in} < s_n > - \sum_{n \epsilon L'_k} J_{kn} < s_n >) \right]. \quad (4.68)$$

Note that in thermal equilibrium,

$$< s_i >_{eq} = \tanh \beta \sum_{k \varepsilon L_i} J_{in} < s_n >_{eq}, \quad (4.69)$$

which, incidentally, is also the equilibrium solution for the single-spin Glauber model in mean field approximation and nearest-neighbour couplings (cf., Eq. 4.53), for $h_i = 0$). This of course is to be expected as both the Glauber and Kawasaki models have the same equilibrium, governed by the underlying Ising Hamiltonian. In order that Eq. (4.69) is ensured, Binder has argued that the primes appearing on the right hand side of Eq. (4.68) in the summations over n must be dropped [101]. In that case it can be easily checked by substituting Eq. (4.69) into Eq. (4.68) that the right hand side of the latter vanishes, if we make use of the identity that

$$\tanh(X - Y) = \frac{\tanh X - \tanh Y}{1 - \tanh X \tanh Y} . \tag{4.70}$$

Hence the Binder prescription makes Eq. (4.68) look like

$$\begin{aligned} 2\lambda^{-1} \frac{d}{dt} < s_i(t) > = & -\xi < s_i > + \sum_{n \epsilon L'_i} < s_k > \\ & + \sum_{k \varepsilon L_i} (1 - < s_i > < s_k >) \\ & \times \tanh \left[\beta (\sum_{k \varepsilon L_i} J_{in} < s_n > - \sum_{k \varepsilon L_i} J_{kn} < s_n >) \right] . \end{aligned} \tag{4.71}$$

We now turn to coarse – grained kinetics, as in the Glauber model. Because we are interested in late-stage dynamics, we approximate, following the decomposition (4.70), as

$$\begin{aligned} & \sum_{k \varepsilon L_i} (1 - < s_i > < s_k >) \tanh \left[\beta (\sum_{k \varepsilon L_i} J_{in} < s_n > - \sum_{k \varepsilon L_k} J_{kn} < s_n >) \right] . \\ & \simeq \tanh \beta \sum_{k \varepsilon L_i} J_{in} < s_n > - \tanh \beta \sum_{k \varepsilon L_k} J_{kn} < s_n > . \end{aligned} \tag{4.72}$$

Substituting in Eq. (4.71) we obtain

$$2\lambda^{-1}\frac{d}{dt} < s_i(t) > = \sum_{k\varepsilon L_i} (< s_k > - < s_i >) -$$

$$\sum_{k\varepsilon L_i}\{\tanh[\beta(\sum_{k\varepsilon L_i} J_{kn} < s_n >] - \tanh[\beta\sum_{k\varepsilon L_i} J_{in} < s_n >]\}$$

$$= \Delta_D\{< s_i > -\tanh[\beta\sum_{k\varepsilon L_i} J_{in} < s_n >]\}, \qquad (4.73)$$

where Δ_D denotes the discrete Laplacian operator. We can now employ the Taylor expansion in Eq. (4.57) to derive the coarse-grained version of Eq. (4.73) as

$$2\lambda^{-1}\frac{\partial m(\vec{r},t)}{\partial t} = -a^2\nabla^2\left[\left(\frac{T_c}{T}-1\right)m - \frac{1}{3}\left(\frac{T_c}{T}\right)^3 m^3 + \frac{T_c}{2T}a^2\nabla^2 m\right], \qquad (4.74)$$

where a is the lattice spacing. Equation (4.74) is of the form of the famous Cahn – Hilliard equation for the phase separation of a binary alloy [102]:

$$\frac{\partial m(\vec{r},t)}{\partial r} = D\,\nabla^2\left(\frac{\delta G[\mathrm{m}]}{\delta m}\right), \qquad (4.75)$$

in which the free energy function G[m] has the structure given in Eq. (4.61) with the local free energy as in Eq. (4.62), and the 'diffusion constant' D is given by

$$D = \frac{\lambda}{2}\beta a^2. \qquad (4.76)$$

The Cahn-Hilliard equation has been very successfully applied to the phase ordering kinetics in a binary AB-alloy, a phase diagram for which is shown in Fig. 4.5. Typically A-A and B-B interactions are attractive ($(\varepsilon^{AA} - \varepsilon^{BB}) < 0$ in Eq. 4.50)), and A-B interactions are repulsive ($-\varepsilon^{AB} > 0$). Hence, it is energetically preferable for the system to separate out into regions rich in A and B, respectively. This is the preferred equilibrium state at low temperatures, i.e. below the co-existence curve in Fig. 4.5. Equation (4.75) then describes the dynamical

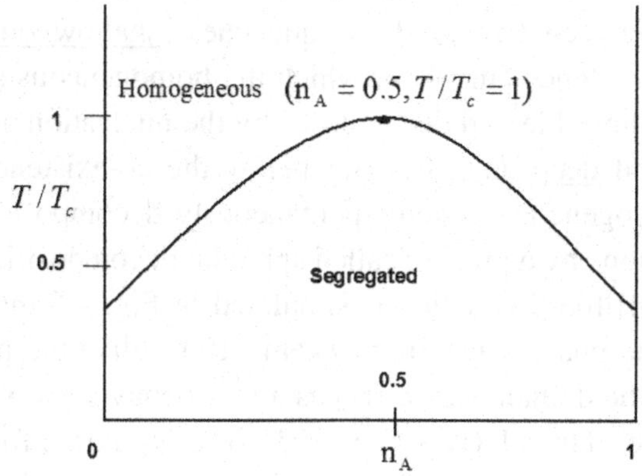

Fig. 4.5. Phase diagram of a binary mixture. The black dot is a second-order critical point.

Fig. 4.6. Simulation of the Cahn-Hilliard model. A-rich regions are marked in black and B-rich regions are unmarked.

evolution due to a quench from above the co-existence curve (homogeneous or disordered phase) to below the co-existence curve (segregated or ordered phase) in Fig. 4.5. In contrast to the nonconserved case, the evolution in this conserved case must satisfy the constraint that numbers of A and B are constant. It is customary to differentiate between two kinds of quenches: shallow quenches (just below the co-existence curve) for which the homogeneous system is not spontaneously unstable and decomposes by the nucleation and growth of the droplets and deep quenches (far below the co-existence curve) for which the homogeneous system spontaneously decomposes into A-rich and B-rich regions by a process called spinodal decomposition.

The Cahn-Hilliard equation is simulated in Fig. 4.6 that depicts the evolution of the phase separating system. But unlike the nonconserved case in which the domain size L (t) has a $t^{1/2}$ behavior, it now exhibits a Lifshitz–Slyozov law: $L(t) \sim t^{1/3}$ [103]. The argument for this, due to Huse [104], runs as follows. The chemical potential on the surface of a domain of size L is $\mu \sim \frac{\sigma}{L}$, where σ is the surface tension. Now, for a conserved system there is an equation of continuity in which the concentration current is given by $D \mid \nabla\mu \mid \sim D\sigma / L^2$, where D is the diffusion co-efficient. Hence, the domain size grows as $\frac{DL}{Dt} \sim \frac{D\sigma}{L^2}$, yielding $L(t) \sim (D\sigma t)^{1/3}$.

Chapter 5

Memory in Nanomagnets

5.1. Introduction to the Physics of Single Domain Nanomagnetic Particles

A bulk ferromagnetic material consists of regions of volume 10^{-12} to 10^{-8} m^3, called domains within which all magnetic moments are aligned. Because each domain consists of 10^{17} – 10^{21} atoms, it may be viewed as a macroscopic system to which the usual laws of thermodynamics apply. The temperature T of the material must be much smaller than the bulk Curie-Weiss temperature T_c in order that all the moments may be orientationally ordered in a given direction. (Note that the temperature T_c is proportional to the exchange coupling J between the individual moments.) The formation of domains is a consequence of minimization of bulk free energy to which come contributions from dipolar interaction in addition to the exchange force, as well as strain couplings. The boundary between different orientations of the magnetization is called a domain wall [105 – 107].

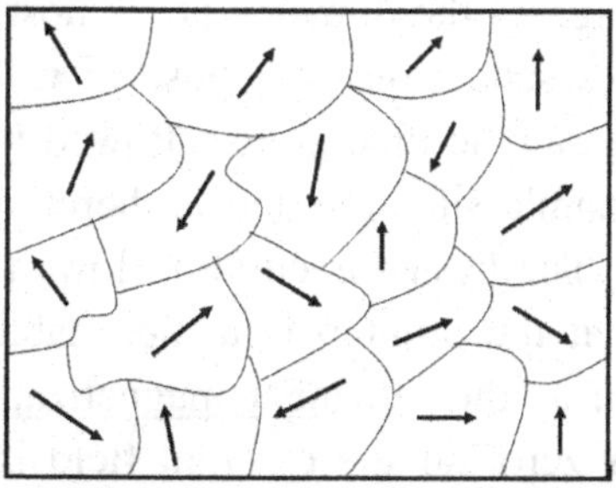

Fig. 5.1. Unmagnetized ferromagnetic specimen.

In the absence of an external magnetic field the domains are randomly oriented so that the total magnetization is zero (see Fig. 5.1). However, when a magnetic field is applied, the domains preferentially align with the field, resulting in a magnetized sample (see Fig. 5.2).

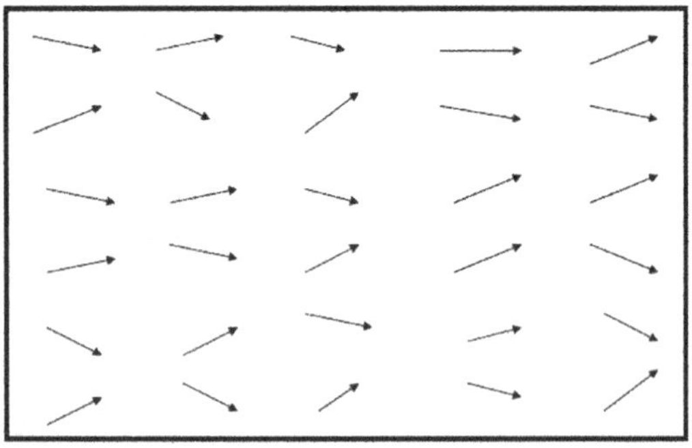

Fig. 5.2. Magnetized ferromagnetic specimen.

Further withdrawal of the magnetic field may not, within measurable times, yield the demagnetized state of Fig. 5.1, but lead to a metastable stable in which the sample may retain a net magnetization in the direction of the magnetic field – a kind of memory effect. This effect is of course temperature–dependent because it is the thermal agitation which is responsible for a phenomenon called Hysteresis. In order to explain hysteresis let us consider the state in Fig. 5.1 in which the total magnetization M is zero for zero applied field H, denoted by the point O in the M-H diagram of Fig. 5.3.

As the field H increases the domains are almost all aligned at a, and the sample approaches saturation. Suppose then the field is suddenly reduced to zero. The sample then is not allowed to immediately adjust to this new themodynamic situation and is therefore led to retain some of its magnetization. The hystereis curve follows the path ab as shown in Fig. 5.3. Note that the point b is a metastable state in which the magnetization, known as the remanent magnetization, is nonzero even though the field H is zero. If the external field is now applied in the reverse direction, the domains reorient until the sample is again

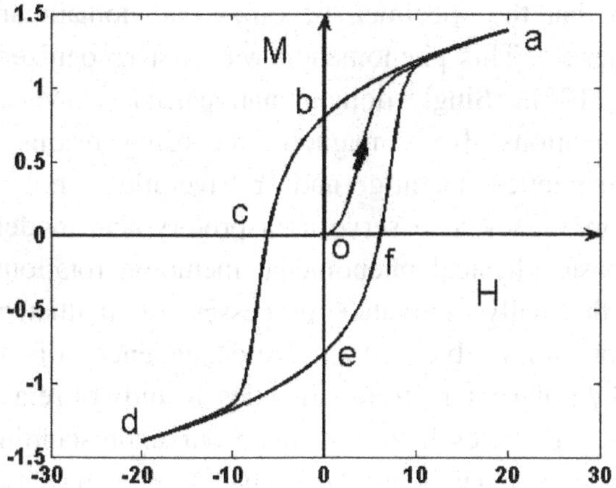

Fig. 5.3. Magnetic Hysteresis.

unmagnetized at point c. When the field is increased further in the reverse direction the sample becomes magnetized in the opposite direction, reaching saturation at point d. Again reversing the direction of the field causes the magnetization curve to follow def; the value of H at f is known as the coercive field. If the field is increased sufficiently, the curve will return to its maximum state of magnetization at a. Clearly the area of the hysteresis loop is given by the closed integral: $\oint MdH$ which is the measure of energy dissipated, implying that the process of hysteresis is an irreversible one. Pretty much the same phenomenon of hysteresis is seen in small ferromagnetic particles though now the non-equilibrium effects are much more pronounced, as argued below.

We now turn attention to the main focus of the present chapter, namely that a ferromagnetic material below a certain critical size (of the order of 20 nm in radius) consists of a single magnetic domain in which a large number of magnetic moments (~10^5 in number) get coherently locked-up in unison. Such a particle may be called a single-domain nanomagnetic particle, the occurrence of which is a consequence of the fact that as the dimensions of the sample are reduced, the surface energty becomes more significant than the volume energy, and there comes a point where it is more favourable energetically to eliminate the domain boundaries so that the specimen becomes one domain and acts as a permanent magnet. This phenomenon was first recognized by Frenkel and Dorfman [105]. Single domain nanoparticles have a remarkable range of applications, from magnetic recording media to catalysis, ferrofluids, magnetic imaging and refrigeration, not to mention paleomagnetism. They also serve as a prototypical model for a wide spectrum of basic physical phenomena, including rotational Brownian motion and thermally activated processes in multistable systems, mesoscopic quantum dynamics, size-dependence of nanomaterial properties and dipolar interactions. Besides, a study of relaxation effects in nanomagnetic particles helps elucidate our understanding of similar phenomena in molecular magnetic clusters, nematic liquid crystals, relaxor ferroelectrics and mictomagnetic-like spin glasses.

Most of the observed effects in single-domain magnetic particles are due to magnetic anisotropy, the energy for which depends on the orientation of the magnetization vector of the domain relative to the crystalline axes. We shall restrict our further discussion to the simplest form of anisotropy which is a uniaxial one (applicable to single-domain particles of Co, say) [107]:

$$E(\theta) = KV \sin^2 \theta, \tag{5.1}$$

where K is the so-called anisotropy energy parameter, V is the volume of the particle and θ is the angle between the magnetization vector and the crystalline anisotropy axis. The origin of K lies in the spin-orbit coupling, acting in conjunction with the electrostatic coupling of the orbit to the crystal [109, 110]. If a field H is applied along the easy axis (E.A.) the energy becomes

$$E(\theta) = KV \sin^2 \theta - \mu VH \cos\theta, \tag{5.2}$$

where μ is the magnetic moment per unit volume. The equilibrium value of magnetization is then given by

$$M_{eq} = V\mu < \cos\theta >_{eq}, \tag{5.3}$$

where

$$< \cos\theta >_{eq} = \frac{\int_0^{\pi} d\theta \sin\theta \cos\theta e^{-\frac{E(\theta)}{K_B T}}}{\int_0^{\pi} d\theta \sin\theta e^{-\frac{E(\theta)}{k_B T}}}. \tag{5.4}$$

For very weak anisotropy (K ~ 0), Eq. (5.4) reduces to the usual Langevin function for classical paramagnets:

$$< \cos\theta >_{eq} \approx \coth\left(\frac{\mu VH}{k_B T}\right) - \frac{k_B T}{\mu VH}. \tag{5.5}$$

On the other hand, if the anisotropy is very large ($KV >> \mu VH$), the magnetic moment is "locked" at the orientation of 0 or π, and hence we may use the 'two-state' limiting formula for:

$$e^{\frac{KV\sin^2\theta}{k_B T}} \sim \frac{1}{2}[\delta(\cos\theta - 1) + \delta(\cos\theta + 1)]. \tag{5.6}$$

In that case

$$\langle cos\theta\rangle_{eq} = \frac{\int_0^{\pi} d\theta sin\theta cos\theta[\delta(cos\theta - 1) + \delta(cos\theta + 1]_e{}^{\frac{\mu HV cos\theta}{k_B T}}}{\int_0^{\pi} d\theta sin\theta[\delta(cos\theta - 1) + \delta(cos\theta + 1]_e{}^{\frac{\mu HV cos\theta}{k_B T}}}$$

$$= \tanh\left(\frac{\mu VH}{k_B T}\right). \tag{5.7}$$

Curiously, in either of these two limits, the average magnetization is independent of the anisotropy parameter K (cf. Eqs. (5.5) and (5.7)).

Note that for low values of applied field $\left(\frac{\mu VH}{k_B T} \ll 1\right)$, Eq. (5.5) yields

$$< \cos\theta >_{eq} \approx \frac{1}{3}\frac{\mu VH}{k_B T}, \tag{5.8}$$

whereas Eq.(5.7) yields

$$< \cos\theta >_{eq} \approx \frac{1}{3}\frac{\mu VH}{k_B T}. \tag{5.9}$$

Thus the magnetization increases by a factor of 3 as the anisotropy parameter K increases from zero to a very large value. On the other hand, the 'saturation' of $< cos\,\theta >_{eq}$, for $\frac{\mu VH}{k_B T} \gg 1$, is unity for both weak and strong anisotropy, as is to be expected.

Until now we have discussed the magnetization properties of single-domain particles when they are in thermal equilibrium. Let us now consider the conditions under which an assembly of (non-interacting) single-domain particles may reach thermal equilibrium in a time short compared with the time of an experiment. One possible route to equilibrium is through physical rotation of particles as can be achieved when they are used as a suspension in a liquid, eg. a 'ferrofluid' consisting of a colloidal suspension of single-domain magnetic iron oxide particles [108]. Clearly, in this mechanism, the factor determining approach to equilibrium is the diffusion coefficient that is related to viscosity of the liquid. The problem at hand is very similar to the one in the context of the Debye treatment of the dielectric properties of a solution containing molecules with permanent electric dipole moments [93]. In the latter case also it is the viscosity of the solution that opposes the rotation of the elementary dipoles.

In a solid, on the other hand, Neel envisaged that for a small enough particle, the total magnetization as a whole, composed of a large number of moments coherently locked-in unison, can undergo rotational relaxation over the anisotropy energy barrier, given by Eq. (5.2) [106]. The corresponding Néel relaxation time (for H = 0) is given by (cf., Fig. 5.4)

$$\tau = \tau_o \exp(KV / k_B T), \tag{5.10}$$

where the pre-exponential factor τ_o is typically of the order of 10^{-9} to 10^{-10} sec. The dependence of the exponent on the volume V of the particle makes τ an extremely sensitive function of the particle-size indeed. For instance, for a spherical iron particle where the only source of anisotropy is its first order crystal anisotropy and K is of the order of 10^{-1} Joule/cm^3, a particle 11.5 nm in radius will have a τ of the order of 10^{-1} sec at room temperature. Such a particle would then reach equilibrium quite rapidly. But a slightly bigger particle of radius 15 nm

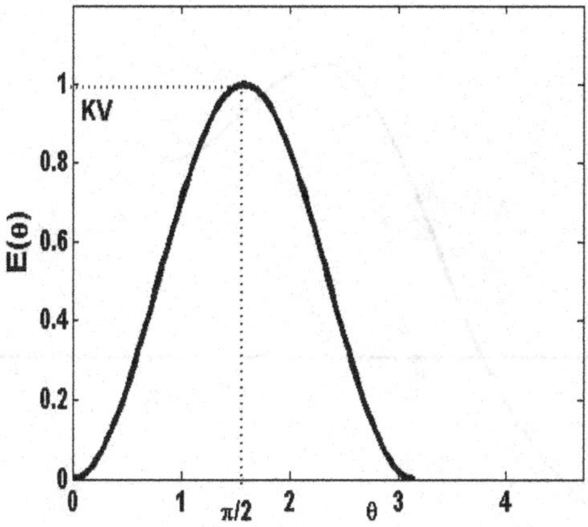

Fig. 5.4. The anisotropy energy barrier: $E(\theta) = KV sin^2\theta$.

will take enormously long time of 10^9 sec to relax and therefore, will be highly stable.

The above discussion brings us then to the issue of the 'time-window' effect in the relaxation behavior of single-domain nano magnetic particles. If $\tau > \tau_{exp,}$ where τ_{exp} is a typical experimental time-window (to be elaborated later) the magnetization vector will appear effectively 'frozen'. On the other hand, for $\tau < \tau_{exp}$, the magnetization will sample both $\theta = 0$ and $\theta = \pi$ configurations equally and several times, thereby yielding a paramagnetic response. Because one is dealing here with not an ordinary paramagnet but a 'giant' moment comprising nearly 10^5 moments in a 'locked-in' orientation, the underlying phenomenon is referred to as "Super-paramagnetic relaxation". In either side of τ_{exp}, an assembly of magnetic nanoparticles will exhibit what is called "magnetic viscosity" and therefore, marked hysteresis effect. The crossover point given by $\tau = \tau_{exp}$ leads to an important concept in the field, viz., that of the "Blocking Temperature" T_B. When $T < T_B$ the moments are frozen, whereas when $T > T_B$ one sees time-dependent 'viscous' effects.

While the above analysis is valid for zero magnetic field the application of an external magnetic field allows for a further tuning of the relaxation time. For a weak magnetic field (compared to the anisotropy energy) the relaxation times are approximated by (cf., Fig. 5.5)

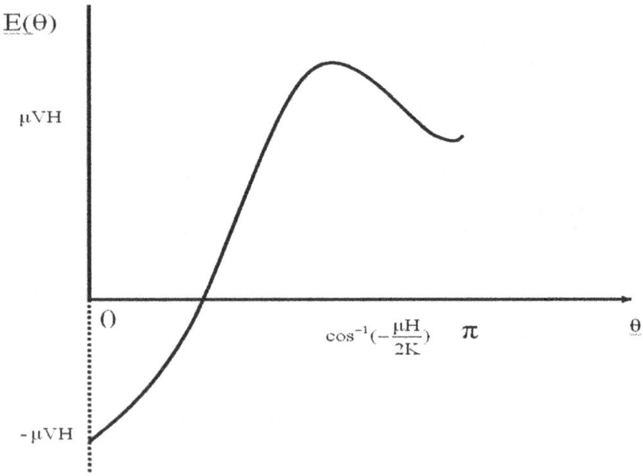

Fig. 5.5. The anisotropy energy barrier in the presence of an applied field H.

$$\tau_{\pm} = \tau_0 \exp(\frac{KV \pm \mu VH}{k_B T}), \tag{5.11}$$

where + corresponds to relaxation out of the $\theta = 0$ orientation whereas – corresponds to relaxation out of the $\theta = \pi$ orientation.

With the preceding background we are in a position to qualitatively discuss the hysteresis curve of Fig. 5.3 which, though mentioned earlier in the context of a bulk ferromagnet with multiple domains, could be equally applicable to a single-domain particle. Starting at the point 0 the relevant energy diagram is given in Fig. 5.4 and the concomitant relaxation time is given by Eq. (5.10). Since both the $\theta = 0$ and $\theta = \pi$ orientations are equally likely we expect to have as many particles with their magnetization 'up' as 'down', in equilibrium, yielding M = 0 for H = 0. Moving on to branch oa, consider an intermediate value of H, say H_1, switched on at time t = 0 (cf., Fig. 5.6). The relevant energy diagram is now given by Fig. 5.5. Relaxation sets in as more and more particles try to have their magnetization 'activated' across the potential barrier to be along the applied field, yielding

$$M(t) = M_{eq}(H_{1,}T)[1-\exp(-t/\tau(V,H_{1,}T))], \qquad (5.12)$$

where $M_{eq}(H_{1,}T)$, the equilibrium value of the magnetization, which can be reached only after waiting for times $t > \tau(V, H_1, T)$, is given by Eqs. (5.5) and (5.7). The relaxation time $\tau(V, H_1, T)$ is obtained from (cf. Eq. (5.11)) as

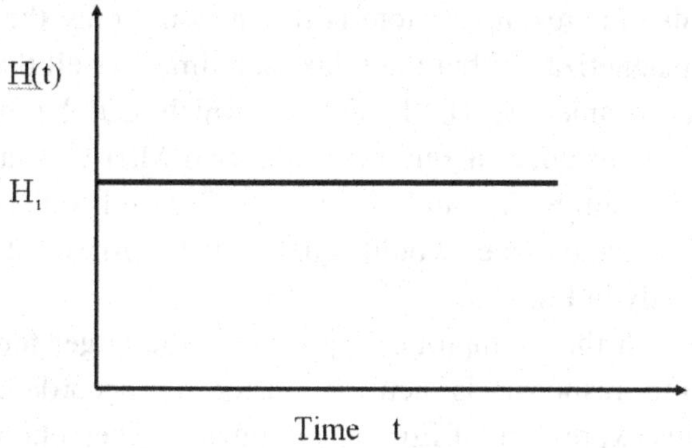

Fig. 5.6. A constant field H_1, switched on, at time $t = 0$.

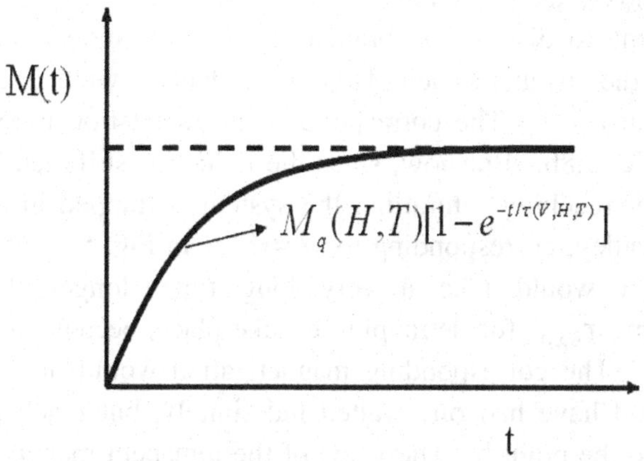

Fig. 5.7. Time-dependent response to the field of Fig. 5.6.

$$\tau^{-1}(V,H_1,T)=\tau_+^{-1}+\tau_-^{-1}. \tag{5.13}$$

The branch a is indeed the 'equilibrium' branch, assuming that $\tau_{exp} > \tau(V,H_1,T)$. Further, if H_1 is weak, $M_{eq}(H_1,T)$ will be given by either Eq. (5.8) or Eq. (5.9), depending on whether the anisotropy parameter K is small or large, respectively. Either way, the value of $M_{eq}(H_1,T)$ will be directly proportional to H_1 and inversely proportional to T, in accordance with the Curie law.

What is also interesting to note is that it is not only the equilibrium value of the magnetization but the relaxation time as well that depend on all the three parameters, H, T and V, which can be tuned in the laboratory. For instance, if one considers two M(t) vs t curves at two different temperatures T_1 and T_2 ($T_2 > T_1$), all other parameters remaining the same, one would qualitatively arrive at a diagram discussed already in Fig. 4.3.

Now although the asymptotic response for the larger temperature T_2 is smaller, the response is actually faster, in accordance with the Néel/Arrhenius expression of Eq. (5.11). Finally, when one increases the field H from H_1 to a sufficiently large value, one reaches saturation (cf., Eq. (5.5) or Eq. (5.7)), for which

$$M_{eq}^{sat} = V\mu, \tag{5.14}$$

which corresponds to the point a in Fig. 5.3.

We now come to discuss the branch ab of the hysteresis curve of Fig. 5.6 when the magnetic field H(t) is suddenly switched off (see Fig. 5.8a), say at $t = \bar{t}$. The corresponding magnetization is expected to relax as in Fig. 5.8b. But now, since the field was sufficiently large to start with, for reaching saturation, the system is trapped in a rather deep potential valley, corresponding to $\theta=0$, as in Fig. 5.5. It is then expected that it would take a very long time, longer than the experimental time τ_{Exp} for detrapping to take place, because of a long relaxation time. The corresponding magnetization would not relax to zero, as it should have had one waited indefinitely, but reach a value corresponding to the point b. The value of the remanent magnetization, given by the length of the segment ob, as indeed the area of the entire hysteresis loop, would of course be temperature-dependent being smaller at a higher temperature, in view of accelerated relaxation (and approach to equilibrium), at a higher temperature.

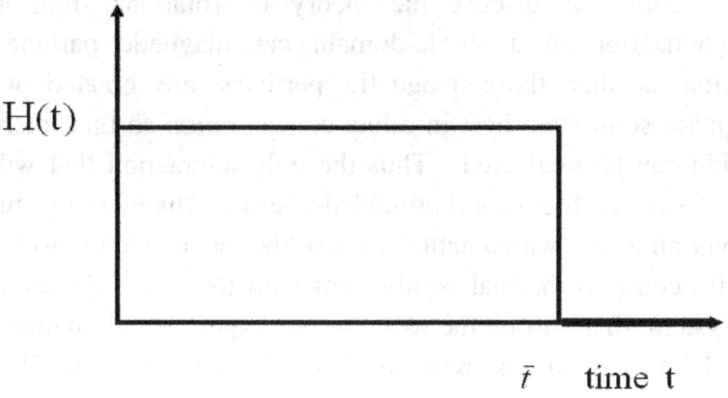

Fig. 5.8a. The field is switched off at $\bar{t}$.

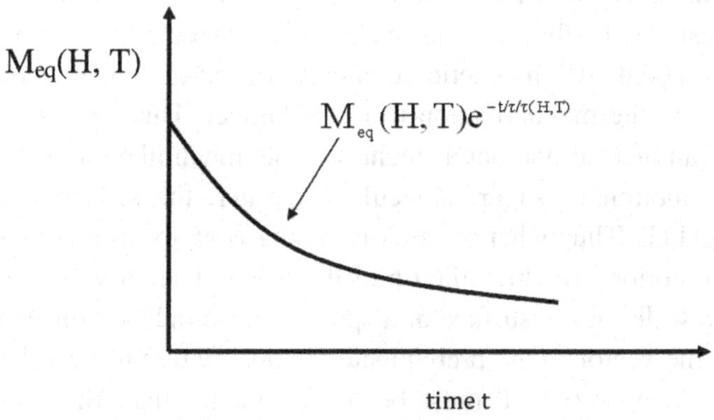

Fig. 5.8b. Relaxation of the magnetization subsequent to $\bar{t}$.

Summing up, the hysteresis phenomenon for single-domain nanomagnetic particles, which can be put to device applications, is a nonequilibrium phenomenon. This phenomenon can be controlled by three distinct tuning parameters, the temperature T, the applied magnetic field H and the volume V of the particle.

5.2. Rotational Brownian Motion, the Kramers Problems and Susceptibility

In this section we discuss the theory of rotational relaxation of the magnetization of a single-domain nanomagnetic particle. Our assumption is that these magnetic particles are created within a nonmagnetic solid state host in dilute concentration so that interparticle interaction can be neglected. Thus the only interaction that will be of concern to us is the heat bath-aided thermal fluctuations, mediated mostly via phonons, which actually cause the magnetization to relax and eventually come to thermal equilibrium with the host. An example of such a system, that will be the focus of our experimental attention, to be discussed later, is an assembly of $NiFe_2O_4$ particles embedded in a nonmagnetic SiO_2 matrix, prepared by the sol-gel technique, in which the ratio of magnetic to nonmagnetic regions can be about 1:6.

Our physical picture of the magnetization-relaxation, borrowed from Néel, is that because the particle is rigidly fixed to the lattice no bodily rotation can be facilitated. Instead, the magnetization vector $\overrightarrow{M}$, that comprises about 10^5 magnetic moments locked-in unison, coherently rotates due to thermal activation from the lattice. This rotation however cannot be an inertial one but a stochastic one, much like the translational Brownian motion of sugar molecules in water, the system treated by Einstein [111]. Thus what we are discussing is an example of rotational Brownian motion, in which the tip of the magnetization vector performs a 'random walk' at the surface of a sphere whose radius (constant) is the length of the vector. One prerequisite for this picture to be valid is that the lattice temperature T must be much smaller than the bulk Curie temperature $T_c (T << T_c)$ for the ferromagnetic specimen (see Fig. 5.1) from which the nanoparticle is created in the first place. This implies that the strong exchange bond (proportional in strength to T_c) between the individual moments <u>within</u> the particle remains solidly intact and we can even ignore small fluctuations of individual moments from the direction of ordering. What we have in hand then is a giant or a 'super' magnetic moment whose value is about 10^5 Bohr magnetons.

The simplest rotational Brownian motion is of course a force (or torque)-free stochastic process given by what is described as a rotational diffusion equation, in total analogy with translational diffusion equation. However, what we are discussing here is Brownian motion in a force field, created by the anisotropy barrier, either in zero magnetic field (Fig. 5.4) or in presence of a magnetic field (Fig. 5.5). Thus the process at hand is a biased random walk in which not all the orientations of the magnetization vector are equally likely to occur. In the sequel we will motivate a rotational diffusion equation in force-field, also known as the Fokker-Planck equation, on the basis of translational Brownian motion in a force-field.

Our analysis is first presented in one dimension which can easily be generalized to three dimensions. Consider the function P(x,t) which measures the probability that a Brownian particle, moving in one dimension, is found at the point x at time t. If we think of the sugar solution then P(x,t)dx can also be interpreted as the sugar concentration within the spatial extent between x and x+dx. As the sugar particles keep moving front one region of the solvent to another, keeping their total number conserved, P(x,t) must obey a continuity equation, reflecting this conservation:

$$\frac{\partial P(x,t)}{\partial t} = -\frac{\partial}{\partial x} j(x,t), \tag{5.15}$$

where j(x,t) is the particle-current. (Note that in dimensions higher than one, j is a vector and the derivative on the right hand side of Eq. (5.15) is a divergence). Now, the motion at hand is a diffusive one for which the current is given by Fick's (or Ohm's Law)

$$j(x,t) = -D\frac{\partial P(x,t)}{\partial x}. \tag{5.16}$$

Combining Eq. (5.15) with Eq. (5.16) we arrive at the diffusion equation:

$$\frac{\partial P(x,t)}{\partial t} = D\frac{\partial^2 P(x,t)}{\partial x^2}. \tag{5.17}$$

While Eq. (5.17) would be quite appropriate for a freely moving Brownian particle and would indeed be the continuum version of a random walk on a one-dimensional discrete lattice, what we would like to consider is the case of Brownian motion in a bias field. The current then would have another contribution, in addition to Eq. (5.16), from the viscous drag of the force-field F(x):

$$j_{force}(x,t) = \frac{F(x)}{\gamma} P(x,t), \tag{5.18}$$

where $\frac{F(x)}{\gamma}$ is the velocity field, γ being the friction coefficient. (Note that for spherical Brownian particles of radius a, γ is related to the viscosity η of the liquid by Stoke's law: $\gamma = 6\pi\eta a$). If the force F(x) is due to a potential E(x), Eq. (5.18) yields

$$j_{force}(x,t) = -\frac{1}{\gamma}\frac{\partial E(x)}{\partial x} P(x,t). \tag{5.19}$$

Adding Eq. (5.19) to Eq. (5.16) we arrive at the Fokker-Planck equation, with the aid of Eq. (5.15):

$$\frac{\partial}{\partial t} P(x,t) = D\frac{\partial^2}{\partial x^2} P(x,t) + \frac{1}{\gamma}\frac{\partial}{\partial x}\left(\frac{\partial E(x)}{\partial x} P(x,t)\right). \tag{5.20}$$

If Eq. (5.20) is to lead to thermal equilibrium, P(x,t) must asymptotically reach the Boltzmann distribution:

$$P_{eq}(x) = \frac{e^{-\frac{1}{k_B T}E(x)}}{\int dx' e^{-\frac{1}{k_B T}E(x')}}, \tag{5.21}$$

where k_B is the Boltzmann constant and T is the temperature. For this to happen we must have the Sutherland-Einstein relation:

$$D = \frac{k_B T}{\gamma} = \frac{k_B T}{6\pi\eta a}. \tag{5.22}$$

The equation (5.20) can then be equivalently expressed as

$$\frac{\partial}{\partial t}P(x,t)=D\frac{\partial}{\partial x}\left[\frac{\partial P(x,t)}{\partial x}+\frac{1}{k_BT}\frac{\partial E(x)}{\partial x}P(x,t)\right]. \tag{5.23}$$

The three-dimensional generalization of Eq. (5.23) is obvious:

$$\frac{\partial}{\partial t}P(x,t)=D\vec{\nabla}_x.\left[D\vec{\nabla}_x P(\vec{x},t)+\nabla\frac{1}{k_BT}(\ _x E(\vec{x}))P(x,t)\right]. \tag{5.24}$$

Our interest here is not of course in the Fokker-Planck equation (5.24) for translational Brownian motion but in the corresponding equation for rotational Brownian motion in a potential-field, shown in either Fig. 5.4 or Fig. 5.5. Since the problem at hand has azimuthal symmetry, in addition to having no time-dependence in spherical polar radius r, the only relevant degree of freedom is the co-latitude angle θ. Writing then all nabla derivative operators in terms of the gradients with respect to θ, Eq. (5.24) can be reduced to

$$\upsilon^{-1}\frac{\partial P(\theta,t)}{\partial t}=\frac{1}{\sin\theta}\frac{\partial}{\partial\theta}\left[\sin\theta\left(\frac{\partial P(\theta,t)}{\partial\theta}+\frac{1}{k_BT}\frac{\partial E(\theta)}{\partial\theta}P(\theta,t)\right)\right],$$

$$0\le\theta\le\pi \tag{5.25}$$

In Eq. (5.25), υ which has the dimension of frequency, can be interpreted as the rotational diffusion constant.

The equation (5.25) was shown by Brown, Jr., to be the basic starting point of studying all relaxation processes involving single-domain magnetic particles [112]. Not only does it yield through the various moments of the probability function $P(\theta,t)$ all the measured magnetic properties but, inter alia, it also leads to a calculational formalism for the Néel relaxation times given by Eqs. (5.10) and (5.11). For instance, the time-dependent response to an applied magnetic field, static or time-independent, would be given by

$$M(t)=\mu V\langle\cos\theta(t)\rangle, \tag{5.26}$$

$$\langle\cos\theta(t)\rangle=\int_o^\pi d\theta\sin\theta\cos\theta P(\theta,t). \tag{5.27}$$

The Zeeman energy due to the applied field (static or time-dependent) would enter Eq.(5.25) through the energy term $E(\theta)$ (cf, Eq. (5.2)), and both linear response and not linear response behavior can be worked out, depending on how the right hand side of Eq. (5.25) is treated [95].

It should be evident to practitioners of nonequilibrium statistical mechanics that the problem we are discussing is a special example of a generic class of problems that were addressed by Kramers [113]. Not only did Kramers provide an (approximate) solution to Eq. (5.25) but in doing so, he also additionally derived the rate of 'escape' over barrier, the inverse of which is related to the relaxation time given by Eq. (5.10). We sketch below the Kramers' treatment. In terms of variable $\xi = \cos\theta$, Eq. (5.25) becomes

$$\upsilon^{-1}\frac{\partial}{\partial t}P(\xi,t) = \frac{\partial}{\partial \xi}\left[(1-\xi^2)\left(\frac{\partial P)}{\partial \xi}\ \frac{1}{k_B T}\frac{\partial E}{\partial \xi}P)\right)\right], \tag{5.28}$$

with $-1 \le \xi \le 1$. The solution of Eq. (5.28) can be written as an eigenfunction expansion:

$$P(\xi,\mathrm{t}) = \sum_n a_n\, \mathrm{e}^{-\lambda_n \mathrm{t}} P_{eq}(\xi) f_n(\xi), \tag{5.29}$$

where $a_n (n \ge 1)$ are coefficients to be determined from the initial conditions, while

$$a_o = 1, \lambda_0 = 0, f_o(\xi) = 1, \lambda_n > 0, \tag{5.30}$$

which guarantee that asymptotically (as $t \to \infty$), $P(\xi,t)$ reaches the thermal equilibrium distribution $P_{eq}(\xi)$. In the Kramers' solution one and only one nonzero eigenvalue is relevant that is associated with the approach to equilibrium. Thus

$$P(\xi,t) = P_{eq}(\xi) + a_1 \mathrm{e}^{-\lambda_n \mathrm{t}} P_{eq}(\xi) f_1(\xi), \tag{5.31}$$

where

$$P_{eq}(\xi) = \frac{e^{-\frac{1}{k_B T}E(\xi)}}{\int d\xi' e^{-\frac{1}{k_B T}E(\xi')}}, \tag{5.32}$$

and for the problem at hand,

$$E(\xi) = KV(1-\xi^2). \tag{5.33}$$

Substituting Eq. (5.31) into Eq. (5.28) the eigenvalue equation can be written as

$$\upsilon \frac{\partial}{\partial \xi}\left[(1-\xi^2)P_{eq}(\xi)\frac{\partial f_1(\xi)}{\partial \xi} + \lambda_1 P_{eq}(\xi) f_1(\xi)\right] = 0. \tag{5.34}$$

The eigenvalue λ_1 can be estimated from the Rayleigh-Ritz inequality, variationally:

$$\lambda_1 \le \upsilon \frac{\int d\xi (1-\xi^2) P_{eq}(\xi)(\frac{\partial \chi}{\partial \xi})^2}{\int d\xi P_{eq}(\xi)\chi^2}, \tag{5.35}$$

where $\chi(\xi)$ is a trial function. The trial function $\chi(\xi)$ must satisfy the normalization condition:

$$\int d\xi P_{eq}(\xi)\chi^2 \xi = N, \tag{5.36}$$

and the orthogonality condition:

$$\int d\xi P_{eq}(\xi)\chi(\xi) = 0, \tag{5.37}$$

N being a constant.

Our task is now to make a suitable choice of the trial function $\chi(\xi)$ in order to estimate λ_1. Referring to Fig. 5.5, since $P_{eq}(\xi)$ is positive in the interval $-1 \le \xi \le 1$, $\frac{d\chi(\xi)}{d\xi}$ must be small in the neighbourhood of the maximum -1 and +1 of $P_{eq}(\xi)$ and peaked around the minimum ξ_m of $P_{eq}(\xi)$ so that the right hand side of Eq.(35) may be minimized. Guided by these considerations, we choose [114]

$$\chi(\xi) = \frac{\chi_1}{1+e^{-a(\xi_m-\xi)}} + \frac{\chi_2}{1+e^{-a(\xi-\xi_m)}}, \tag{5.38}$$

where χ_1 and χ_2 are constants and a is a variational parameter. It is evident that for suitably large a, $\chi(\xi)$ should rapidly approach $\chi_1(\chi_2)$ for $\xi < \xi_m (> \xi_m)$ and hence χ_1 and ξ_2 must be of opposite signs, in accordance with Eq. (5.37). In addition, the derivative of χ, viz., becomes sharply peaked around $\xi = \xi_m$ and rapidly falls off to zero on both sides of $\xi_{m.}$ Dimensional reasoning then suggests that we might express a in the form.

$$\frac{d\chi}{d\xi} = \frac{a}{2}(\chi_2 - \chi_1)(1 + \cosh a(\chi_m - \chi))^{-1}, \tag{5.39}$$

$$a^2 = b|E''(\xi_m)|/k_B T, \tag{5.40}$$

where b is a new variational parameter and $|E''(\xi_m)|$ is the curvature of the potential at its maximum. The sharp peaking argument then tacitly implies that the ratio of the curvature of the potential at its maximum to the thermal energy $k_B T$ must be suitably large. The minimization procedure, as sketched above, finally yields for the eigenvalue λ_1 the Kramers expression which we quote her for H = 0, appropriate to Fig. 5.4 [87]:

$$\lambda_1 = 4\pi\upsilon\left(\frac{KV}{\pi k_B T}\right)^{3/2} \exp\left(-\frac{KV}{k_B T}\right). \tag{5.41}$$

The inverse of λ_1 then leads to the Néel formula (Eq. (5.10) in which τ_0 can be easily read off from Eq. (5.41).

With the preceding machinery at hand we return to Fig. 5.6 and consider the linear response to a uniform but small magnetic field H, applied from the time t = 0 onwards. This response is characterized, in linear response theory, by the response function [87].

$$\Psi(t) \equiv \lim_{H\to 0} \frac{M(t)}{H} = \frac{1}{k_B T}(\langle M^2\rangle_0 - \langle M(0)M(t)\rangle_0). \tag{5.42}$$

The crucial feature of Eq. (5.42) is that the fluctuating quantities e.g., the mean-squared magnetization, given by $\langle M^2\rangle_{0,}$ and the time auto-correlation function of the magnetization, given by $\langle M(0)M(t)\rangle_{0,}$ are calculated in the absence of external perturbation, viz., the magnetic field H, when the system was in thermal equilibrium (for instance, for t < 0 in Fig. 5.6). This feature is underscored by inserting a subscript zero underneath the expressions on the right hand side of Eq. (5.42).

The Fokker-Planck equation (5.25) is eminently suitable for not only calculating nonequilibrium properties as in Eq. (5.27) but evaluating fluctuating quantities in thermal equilibrium as well. Thus the correlation function appearing in Eq. (5.42) is given by

$$\langle M(0)M(t)\rangle_0 = \mu^2 V^2 \iint \cos\theta_0 P_{eq}(\theta_0) P(\theta,t) \cos\theta \sin\theta_0 d\theta_0 \sin\theta d\theta. \tag{5.43}$$

Evidently, $P(\theta,t)$ must satisfy the initial condition

$$P(\theta, t=0) = \delta(\cos\theta - \cos\theta_0). \tag{5.44}$$

The appearance of $P_{eq}(\theta_0)$ is in accordance with the fact that the system is in thermal equilibrium, initially at t = 0. It is interesting to note that Eq. (5.30), together with Eq. (5.44), automatically yields the static fluctuation, upon setting t = 0:

$$\langle M^2\rangle_0 = \mu^2 V^2 \int \cos^2\theta_0 P_{eq}(\theta_0) \sin\theta_0 d\theta_0. \tag{5.45}$$

In order to compute the right hand side of Eq. (5.43) we need the Kramers solution given by Eq. (5.31). But for the latter to be compatible with the initial condition in Eq. (5.44) we need to rewrite Eq. (5.31) as

$$P(\theta,t) = P_{eq}(\theta) + [\delta(\cos\theta - \cos\theta_0) - P_{eq}(\theta)]e^{-\lambda_{KR}t}, \tag{5.46}$$

where we have inserted a subscript KR under λ, in deference to Kramers. This equation then properly incorporates both the initial condition (5.44) and the asymptotic property (viz., that $\lim_{t\to\infty} P(\theta,t) = P_{eq}(\theta)$).

Substituting Eq. (5.46) into Eq. (5.43) yields

$$\langle M(o)M(t)\rangle_0 = \langle M^2\rangle_0 exp\,(-\lambda_{KR}t), \tag{5.47}$$

since, in the present example of a symmetric potential (cf. Fig. 5.4), for which both the $\theta = 0$ and $\theta = \pi$ orientations of magnetization are equally probable in equilibrium,

$$\int \cos\theta P_{eq}(\theta)\sin\theta d\theta = 0 \tag{5.48}$$

The Eq. (5.48) is of course a statement of the fact that the magnetization in equilibrium is zero, as in the present case. The response function in Eq. (5.42) is then given by

$$\Psi(t) = \frac{< M^2 >_0}{k_B T}(1 - e{-}\lambda_{KR}t). \tag{5.49}$$

We have thus come a full circle because we have just provided a derivation of Eq. (5.41) which was motivated earlier from physical considerations. Therefore, comparison of Eq. (5.49) with Eq. (5.41) leads to an identification of the Kramers rate λ_{KR} with the inverse of the relaxation time τ, and additionally,

$$M_{eq}(H,T) = \frac{H}{k_B T}\langle M^2 \rangle_0, \tag{5.50}$$

recalling that $M(t) = H\Psi(t)$, in linear response theory. The Eq. (5.50) is of course the well known expression for the equilibrium susceptibility $(\chi_{eq} \equiv \frac{M_{eq}(H,t)}{H})$ in terms of spontaneous magnetization fluctuations, given by Gibbsian statistical mechanics [115]. It is tempting also to check under what condition does the right hand side of Eq. (5.50) agree with the expression derived under Eq. (5.8) or Eq. (5.9) earlier. Note that from Eq. (5.45),

$$< M^2 >_0 = \mu^2 V^2 \int_0^{\pi} d\theta_0 \sin\theta_0 \cos^2\theta_0 e^{\frac{KV Sin^2\theta_0}{k_B T}}$$

$$= \mu^2 V^2 \int_{-1}^{1} dx x^2 e^{\frac{KV}{K_B T}(1-x^2)}. \tag{5.51}$$

For large anisotropy $(KV >> k_B T)$, which is in fact the regime in which the Kramers treatment is valid (see the sentence following Eq. (5.40)), the exponential under the integral in Eq. (5.51) can be written as the sum of two delta functions, as argued earlier (see Eq. (5.6)). Then,

$$\langle M^2 \rangle_{\overline{0}} = \mu^2 V^2, \tag{5.52}$$

where we have ignored terms of $O\left({k_B T}/{KV}\right)$.

Substituting in Eq. (5.50), we find

$$M_{eq}(H,T) = \frac{\mu^2 V^2 H}{k_B T}, \tag{5.53}$$

which agrees with the combination of Eq. (5.3) and Eq. (5.9), underscoring once again the fact that Kramers' is a 'high barrier-weak noise' theory (i.e. $KV / k_B T >> 1$).

Even though we considered above the response to a static magnetic field the remarkable feature of linear response theory is that the dynamic (but asymptotic) response to a weak frequency – dependent magnetic field can also be directly computed from the response function $\Psi(t)$ [87]. Thus the frequency–dependent susceptibility is given by

$$\chi(\omega) = \lim_{\delta \to 0} s\Psi(s), \tag{5.54}$$

where $\Psi(s)$ is the Laplace transform of $\Psi(s)$, defined by

$$\Psi(s) = \int_0^\infty dt e^{-st} \Psi(t), \tag{5.55}$$

in which the transform variable is defined by

$$s = \delta - i\omega \tag{5.56}$$

Thus

$$\chi(\omega) = \frac{\langle M^2 \rangle_0}{k_B T} \frac{\lambda_{KR}}{\lambda_{KR} - i\omega}. \tag{5.57}$$

Combining with Eq. (5.52) and writing λ_{KR} as the inverse of the relaxation time τ, we have

$$\chi(\omega) = \frac{\mu^2 V^2}{k_B T} \frac{1}{1 - i\omega\tau}. \tag{5.58}$$

Equation (5.58) deserves a few remarks:

(i) The frequency–dependent response is characterized by a single parameter, the relaxation time τ, which goes under the name of 'Debye response'. This identification once again emphasizes the main theme of these Lecture Notes, namely the influence of the subject of Magnetism in other areas. In this case we find that relaxation of single-domain magnetic particles has much in common with electric dipole relaxation in polar liquids. Indeed, the analogy is stronger in the case of Ferrofluids in which nano-magnetic particles are dispersed in a liquid. (ii) It is clear that ω^{-1} provides an experimental time-window. When this window is very wide, i.e. $\omega\tau \ll 1$, the response is given by static (real) susceptibility

$$\chi_0 = \chi(\omega = 0) = \frac{\mu^2 V^2}{k_B T}. \tag{5.59}$$

This has the characteristic $(k_B T)^{-1}$ temperature dependence – the Curie law. The above result is easy to understand: when a nanomagnetic particle is probed at a time scale much larger than the relaxation time τ, the particle moment would reorient many times over and behave exactly like a paramagnet. On the other hand, when the two time scales are compatible i.e. $\omega\tau = 1$, the response is marked by strong frequency – dependent effects, familiar in Rock Magnetism [116]. (iii) The imaginary part of the susceptibility, which characterizes dissipation in the system, is given by

$$\chi''(\omega) = \frac{\mu^2 V^2}{k_B T} \frac{\omega\tau}{1 + \omega^2\tau^2}. \tag{5.60}$$

If we plot $\chi''(\omega)$ versus ω we obtain a symmetric peak, called the Debye peak with a width given by

$$\Gamma = {}^{2/3}/_{\tau}. \tag{5.61}$$

Hence a measurement of the width yields τ, from which the barrier height can be estimated (cf. Eq. 5.41)) in a semilog plot of τ against the temperature T.

(iv) The real part of the susceptibility is

$$\chi'(\omega) = \frac{\chi_0}{1+\omega^2\tau^2}. \tag{5.62}$$

Using the expression for τ given by Eq. (5.10) and dividing $\chi'(\omega)$ by χ_0 at T = 300K (cf., Eq. (5.59)), we have earlier shown in Fig. 4.2 the graphs of $\chi'(\omega)$ versus T, for two values of ω.

The typical hump is characteristic of 'thermal freezing' of the magnetization and occurs at a temperature called the 'Blocking Temperature' T_B which is also a function of the frequency ω. Thus T_B is a property of the experimental time–window: $T > T_B$ leads to paramagnetic response which is Curie like (T^{-1}) whereas $T < T_B$ blocks the magnetization. Polydispersity (i.e., size distribution) will of course cause a broadening of the hump.

5.3. Experiments: Stern–Gerlach and Mössbauer Spectroscopy

With the introductory background to the phenomenology of single-domain nano-magnetic particles and the underlying relaxation theory we turn our attention in this section to experimental measurement of the relevant properties. The parameters of interest which can be extracted from experiments are:

a) size or volume V of the particles;
b) the magnetic moment per unit volume μ;
c) the anisotropy energy K, which provides the barrier to relaxation;
d) polydispersity, viz., the distribution function that characterizes the size-distribution, if there is any;
e) and interparticle interaction, the most commonly encountered one being the dipolar one.

We have so far ignored (e) and considered only independent, non-interacting magnetic particles, and the only measurement techniques we have touched upon in passing are the magnetization and the susceptibility. In what follows, we shall examine a variant of the celebrated Stern-Gerlach setup that has been recently used to study relaxation of nanomagnetic particles, especially magnetic clusters [117].

Magnetic Clusters in a Stern-Gerlach Setup

The experimental configuration is shown in Fig. 5.9. The only difference with the historic experiment is what are jettisoned from the oven are not singly charged Ag ions but single-domain clusters of Co, Gd, etc. For a cluster whose magnetization μV is polarized along the direction z of the inhomogeneous magnetic field one would have expected in the old measurement that the deflection d on the detector screen, measured from $z = 0$, would be given by

$$d = \frac{\mu V}{M} B'(o) \frac{LD}{v^2} (1 + \frac{L}{2D}), \tag{5.63}$$

where m is the mass of the particle, $B'(o)$ is the gradient of the magnetic field B(z) at $z = 0$, v is the (constant) speed with which the jet emanates from the oven in the forward (i.e. normal to z-axis) direction and the Length L and D are defined in Fig. 5.9.

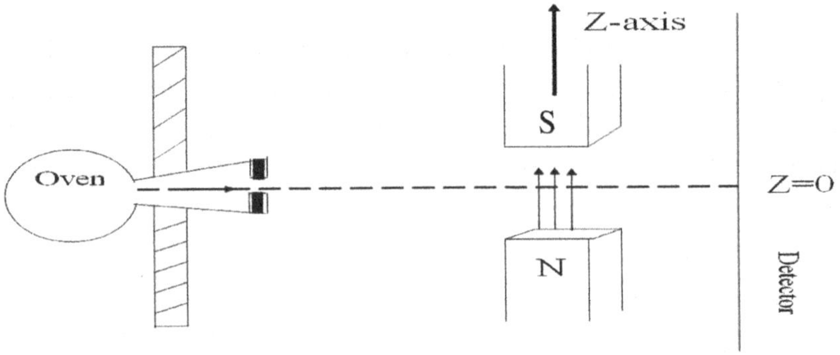

Fig. 5.9. Stern-Gerlach Setup.

The first contribution within the round brackets in Eq. (5.63) comes from ballistic motion outside the magnet while the second term arises from the accelerated motion within the magnet. Though Eq. (5.63) is wrong in the present discussion because of rotational relaxation of the cluster magnetization within the magnetic pole-pieces region, as will be shown below, it allows for the introduction of an important parameter τ_E that defines the experimental time-window for the setup at hand:

$$\tau_E \equiv \frac{L}{\mathrm{v}}. \tag{5.64}$$

The reason Eq. (5.63) is wrong is that the velocity $\mathrm{v_z}$ of the cluster, derived from the acceleration gained in the z-direction in view of the force exerted by the inhomogeneous magnet on the magnetic moment of the cluster, is now a stochastic process, given by

$$\mathrm{v_z}(t) = \alpha \int_0^t \cos\theta(t)dt, \tag{5.65}$$

where α is a constant parameter of the experiment: $\alpha \equiv \frac{\mu V}{M} B'(o)$ and $\theta(t)$ is the angle between the magnetic field (i.e. z-axis) and the magnetization vector of the particle. For reasons discussed in detail in Sec. 5.2 the relaxational dynamics of $\cos\theta(t)$ may be described in terms of the Fokker-Planck equation (5.25).

We would like to dispose of one red herring in order to facilitate comparison with experimental data. Note that the initial conditions chosen on $\mathrm{v_z}(t)$ is that it is zero at time $t=0$, implying that the molecular jet is unidirectional. However, while this is predominantly so, there is a small residual fluctuation in the velocity in the z-direction due to the temperature T_s of the source i.e. the oven. That is, over and above the very large speed of the jet along the dashed line in Fig. 5.9, there is an envelope of velocity distribution governed by a Maxwellian at temperature T_s, thus

$$\mathrm{v_z}(t) = \Delta\mathrm{v_z} + \alpha \int_0^t \cos\theta(t)dt, \tag{5.66}$$

where

$$<(\Delta\mathrm{v}_z)^2> = \frac{k_B T_s}{m}. \tag{5.67}$$

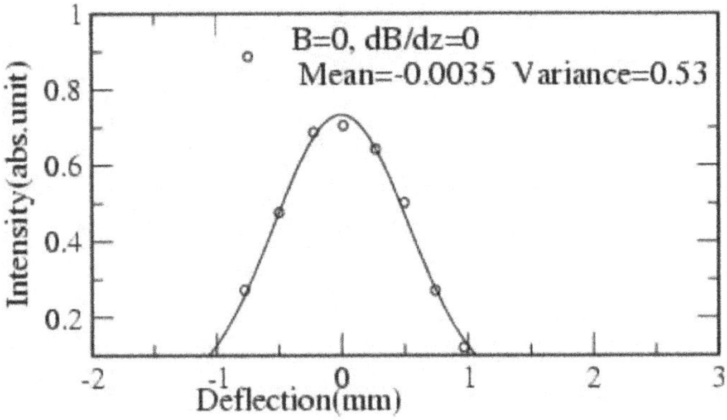

Fig. 5.10. Line profile, even at zero field, implying that there is a transverse fluctuation in velocity.

This fact is borne out in the experimental data shown in Fig. 5.10 which indicates that even for zero magnetic field (and its gradient) there is sizable broadening of the deflection profiles. This residual broadening, established from Eq. (5.67), is then added to all the theoretically computed line profile below, for a meaningful comparison with data. This discussion then also brings to fore the existence of two distinct temperatures in the problem: the oven temperature T_v already mentioned, which determines the jet-speed, and the lattice or the vibration temperature T_υ that governs the rotational relaxation of the magnetization of the cluster, as it tries to equilibrate to its own lattice through spin-lattice relaxations. It is T_v that enters the discussion of the Fokker – Planck equation (5.25). (For a more detailed discussion on T_s and T_v see [117]).

With the preceding background to the Stern-Gerlach geometry, the deflection d on the detector screen, measured from $z=0$ (and upon ignoring Δv_z in Eq. (5.65)) can be written as

$$d = \frac{\alpha D}{\mathrm{v}} \int_0^{\tau_E} \cos\theta(t')dt' + \alpha \int_0^{\tau_E} dt \int_0^{t} \cos\theta(t')dt'. \tag{5.68}$$

Clearly, d itself is a <u>driven</u> stochastic process, the mean value of which has to be computed from the expression (5.27), and the Kramers' solution for the probability function $P(\theta,t)$. We have found earlier that [118]

$$\bar{d} = \frac{\alpha D}{\alpha} p_0 \left\{\frac{1}{\lambda_{KR}}\left(1 - e^{-\lambda_{KR}\tau_E}\right) + \frac{\alpha}{D\lambda_{KR}}\left[\tau_E - \frac{1}{\lambda_{KR}}\left(1 - e^{-\lambda_{KR}\tau_E}\right)\right]\right\}$$

$$+\alpha < \cos\theta >_{eq} \left\{\frac{1}{2}\tau_E^2 + \frac{D}{\mathrm{v}}\left(1 - \frac{\mathrm{v}}{D\lambda_{KR}}\right)\left[\tau_E - \frac{1}{\lambda_{KR}}\left(1 - e^{-\lambda_{KR}\tau_E}\right)\right]\right\}. \tag{5.69}$$

where λ_{KR} is the Kramers' relaxation rate that depends on the barrier height (anisotropy energy) and also the possible presence of a homogeneous component of the magnetic field in the Stern-Gerlach setup, $< \cos\theta >_{eq}$ is given by Eq. (5.4) and p_0 is the initial polarization of the entering jet. Note that in the static limit ($\lambda_{KR} = 0$), applicable to the original Stern-Gerlach experiment, and for a fully polarized beam ($p_0 = 1$), Eq. (5.69) reduces to Eq. (5.63), as we had anticipated earlier.

The first step towards calculating the line profile, i.e., the intensity of 'hits' on the detector versus the deflection, requires the variance of the deflection. For computing the latter we want to evaluate the correlation function: $< \cos\theta(0)\cos\theta(t) >$, which has been the subject of our earlier discussion (cf., Eq. (5.43)). But, now, $< \cos\theta >_{eq}$ is nonzero. We find

$$< cos\theta(0)\cos\theta(t) >$$

$$=< cos^2\theta_0 > e^{-\lambda_{KR}t} + p_0 < cos\theta >_{eq} \left(1 - e^{-\lambda_{KR}t}\right). \tag{5.70}$$

The Eq. (5.70) is needed to compute $\overline{d^2}$ and from that, the variance $\left(\overline{d^2} - \bar{d}^2\right)$, employing Eq. (5.69). While the variance is just the second comulant of the probability distribution of the deflection $\pi(d)$, in principle, all higher order cumulants are required to calculate the intensity profile. We find incredibly though that the second cumulant suffices i.e. the probability function $\pi(d)$ is nearly a Gaussian [119]. The comparison with experimental data for Gd, and Co clusters is shown in Fig. 5.11 and Fig. 5.12 respectively. Figure 5.13 is a depiction of the intensity profile for Gd_{22} clusters which seems to point to a bidisperse

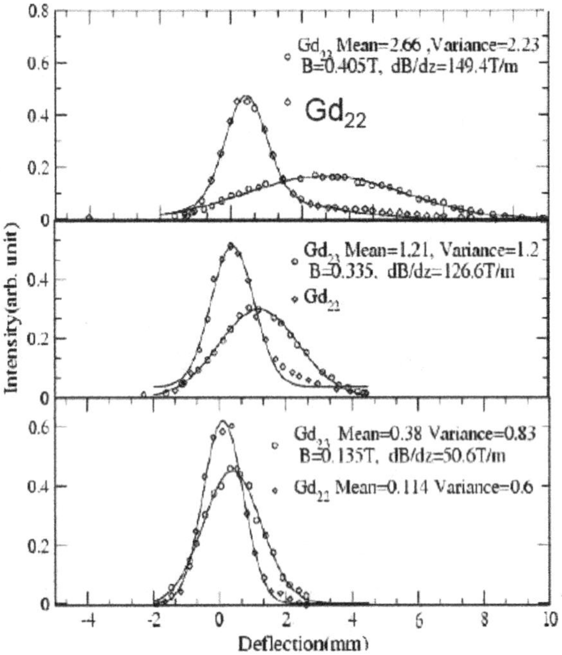

Fig. 5.11. Line profile for Gd_{23} and Gd_{22} clusters.

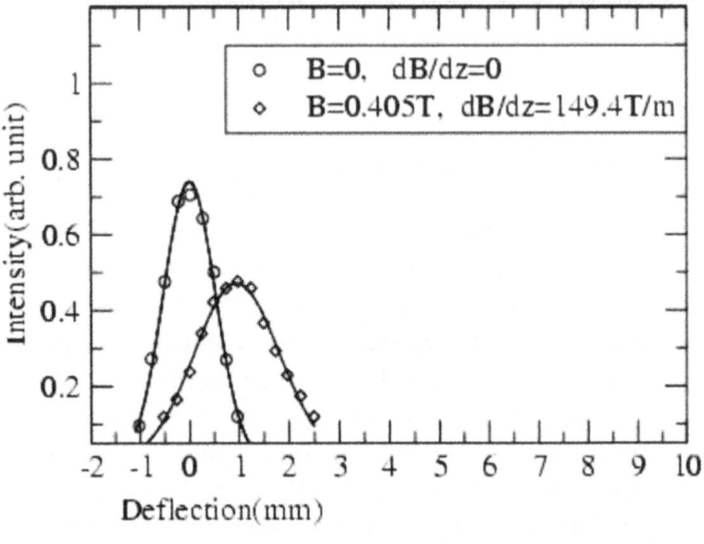

Fig. 5.12. Line profile for co-clusters, in zero and finite magnetic fields.

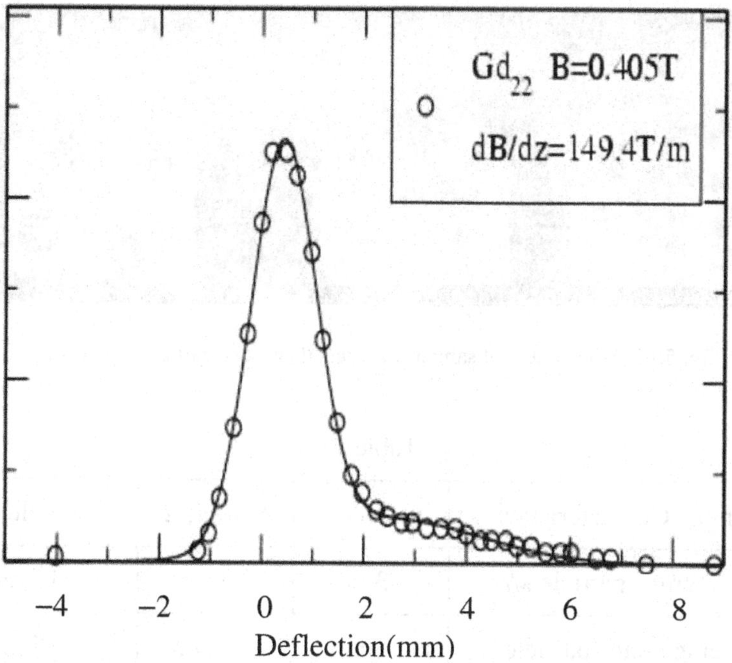

Fig. 5.13. Double Gaussian profile for Gd_{22} indicating that there is a bi-modal size distribution.

distribution of sizes in the sample concerned. Thus we have a superposition of a set of clusters whose magnetization is frozen $(\tau > \tau_E)$ and another set that exhibits superparamagnetism $(\tau < \tau_E)$.

Mössbauer Relaxation of $NiFe_2O_4$ in SiO_2 Matrix

In this subsection we describe Mössbauer spectroscopy of single-domain nanomagnetic particles of $NiFe_2O_4$ embedded in a nonmagnetic SiO_2 host. The samples are prepared by standard sol-gel techniques and are characterized by X-ray diffraction (XRD) and transmission electron microscopy (TEM) [120]. A rough description of three different kinds of samples used is shown in Fig. 5.14. The Table 1 below provides the relevant parameters:

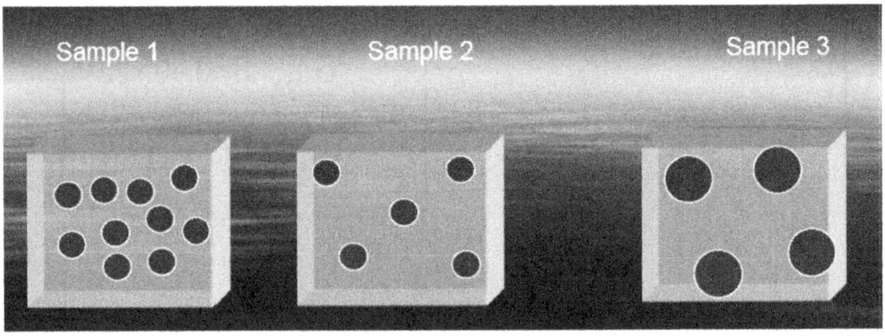

Fig. 5.14. A schematic of sample sizes and their inter-particle separation.

Table 1

Sample Characteristics	Sample 1	Sample 2	Sample 3
Average particle size	3 nm	3 nm	15 nm
Average inter-particle distance	5 nm	15 nm	15 nm

In all cases the particle-size is found to follow a long-normal distribution which has the known attribute that dispersion goes hand in hand with the average size. Thus sample 3 with average particle size 15 nm is also expected to have the most pronounced dispersion compared to sample 1 and 2. We estimate that for sample 1, the inter-particle dipolar interaction is the strongest because the average separation is about 5 nm. On the other hand, dipolar coupling is indeed very weak, for samples 2 and 3. Thus, sample 1 will be notionally referred to as "Interacting" while samples 2 and 3 as "Noninteracting". This nomenclature is borne out by the room temperature hysteresis data, shown in Fig. 5.15. At room temperature the Neel relaxation is expected to be extremely rapid (cf. Eq. (5.11)) but the difference in the hysteresis curves for samples 2 and 3, in which sample 2 shows no hysteresis but sample 3 exhibits weak hysteresis, is precisely due to the size difference.

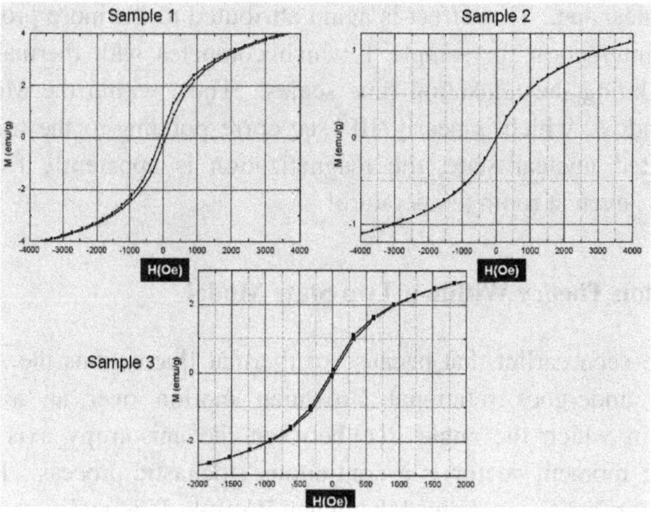

Fig. 5.15. Hysteresis behavior for samples 1, 2 and 3.

The 15 nm sized sample 3 has its relaxation time somewhat elongated (cf. Eq. (5.11)) compared to the 3 nm sized sample 2, and therefore, sample 3 requires a somewhat longer time to reach equilibrium. On the other hand, there is moderate hysteresis for sample 1 even though the average particle size ($\simeq 3\ nm$) is the same as for sample 2; the reason is the dipolar interaction. For sample 1, the average interparticle separation is 5 nm and hence, the dipolar coupling, which is inversely proportional to the cube of inter-particle distance, is estimated to be stronger than in sample 2, again slowing down relaxation.

In order to focus onto the effect of inter-particle interaction, keeping the average size fixed, we show temperature-dependent Mössbauer spectra in Fig. 5.16 for sample 1, and Fig. 5.17 for sample 2. The data at 20 K for both samples exhibit the characteristic six-finger pattern for 57_{Fe}, contained in $NiFe_2O_4$, although sample 2 displays slight broadening of lines, implying that relaxation is not completely frozen, even at 20 K. As the temperature increases, relaxation speeds up and for sample 2 and at room temperature (RT), complete 'motional narrowing' sets in, yielding a single absorption line. Evidently, this speeding up of relaxation is not adequate for sample 1 which, even at RT, demonstrates a split-lineshape. This effect is again attributed to the more pronounced dipolar interaction for sample 1 which competes with thermal energy in modulating the relaxation time scales. Thus, within the Mössbauer time window, which is nearly 10^{-8} sec corresponding to the lifetime of the excited nuclear state, the magnetization is apparently frozen for sample 1, even at room temperature!

Relaxation Theory Within a Two State Model

We have seen earlier that because of thermal fluctuations the magnetic moment undergoes rotational Brownian motion over an anisotropy barrier, in which the angle $\theta(t)$ between the anisotropy axis and the magnetic moment vector is a continuous stochastic process. However the Kramers' analysis showed that when $KV \gg k_BT$ (in reality though KV can be a few times larger than k_BT in order that the Kramers analysis is valid), the magnetic moment is mostly locked-in two orientations $\theta = 0$ and $\theta = \pi$, with slow relaxation between the two configurations. We are thus in the so called "Ising' limit in which $\theta(t)$may be viewed as a dichotomic Markov process, in which it jumps at random between the angels 0 and π at a rate given by the Arrhenius – Kramers formula (Eq. 5.10) (cf., also Fig. 4.1).

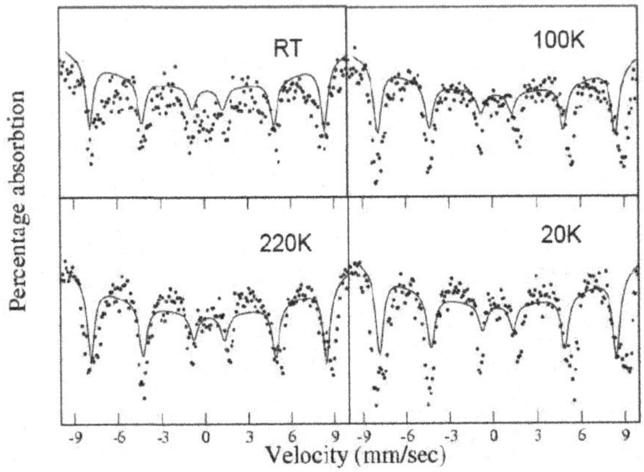

Fig. 5.16. Mössbauer lineshape for sample 1. Even RT spectre show split lines.

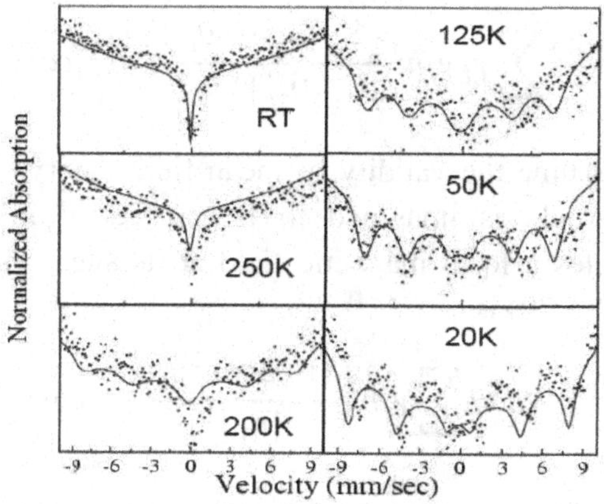

Fig. 5.17. Mӧssbauer spectre for sample 2. The RT spectrum indicates 'motional narrowing'.

We now discuss in somewhat more detail the effect of dipolar interaction. The latter is well known to yield complex magnetic order, being long-ranged, anisotropic and alternating in the sign of interaction. However, our hysteresis measurement, exhibited in Fig. 5.15, indicates that even for sample 1 for which the dipolar coupling is relevant, there is no shift of the hysteresis loop, thereby implying that the bulk magnetization, for the zero applied field, is zero. Our interpretation is that because of the largeness of the anisotropy barrier ($\propto K$) we are in the Ising limit of the dipolar interaction, for which the local field, in the mean field sense, points parallel or antiparallel to the anisotropy axis, with equal probability. The dipolar interaction can now be described by its truncated form:

$$\mathcal{H}_{d-d} = \sum_{ij} \gamma_i \gamma_j \hbar^2 \frac{(1 - 3cos^2\theta_{ij})}{|\vec{r_{ij}}|^3} m_{zi} m_{zj}, \tag{5.71}$$

where $\gamma_i$and γ_j are the gyromagnetic ratio of the ith and j^th^ particles respectively, $\overrightarrow{r_{ij}}$ is the vector distance between the sites at which the particles are located, $\theta_{ij}$is the angle between the $\overrightarrow{r_{ij}}$ and the anisotropy axis and m_{zi} is the (giant) magnetic moment for the ith nanoparticle along the anisotropy axis (i.e. Z). Given that m $\propto$ V, Eq. (71) may be rewritten as

$$\mathcal{H}_{d-d} = \mu^2 V^2 \sum_{ij} \gamma_i \gamma_j \hbar^2 \frac{(1 - 3cos^2\theta_{ij})}{|\overrightarrow{r_{ij}}|^3} cos\theta_i cos\theta_j. \qquad (5.72)$$

Further, we assume the validity of mean field theory in which each nanomagnetic particle is imagined to be embedded in an effective medium that creates a local magnetic field at its site. The mean field (MF) form of Eq. (5.72) is

$$\mathcal{H}_{d-d}^{MF} = \gamma\hbar\mu^2 V^2 cos\theta \sum_{j} \gamma_j \hbar \frac{(1 - 3cos^2\theta_{ij})}{|\overrightarrow{r_{ij}}|^3} < cos\theta_j >, \qquad (5.73)$$

where the angular brackets < …. > represent a thermal average. Further, in accordance with our assumption about the largeness of the anisotropy energy, $cos\theta$ can be replaced by a two-state Ising variable σ:

$$\mathcal{H}_{d-d}^{MF} = \gamma\hbar\mu^2 V^2 \sigma \sum_{j} \gamma_j \hbar \frac{(1 - 3cos^2\theta_{ij})}{|\overrightarrow{r_{ij}}|^3} < \sigma_j >. \qquad (5.74)$$

In line with this approximation each particle can be viewed to be subjected to a local magnetic field H such that

$$\mathcal{H}_{d-d}^{MF} = \mu V \sigma H, \qquad (5.75)$$

$$H = \mu \Lambda V < \sigma >, \qquad (5.76)$$

where Λ is a parameter that subsumes all the other constants. Within the proposed mean field theory, H can be expressed as (cf., Eq. (5.7))

$$H = \mu \Lambda V \tanh\left(\frac{\mu V H}{k_B T}\right). \qquad (5.77)$$

Note that Eq. (5.77) admits both positive and negative solutions for H, in accordance with out discussion preceding Eq. (5.71).

We now turn our attention to relaxation kinetics. Within the previously stated physical picture in which the Fokker-Planck process is replaced by a dichotomic Markov process we may write down the following set of rate equations for the number of magnetic particles with a specific orientations of their magnetization:

$$\frac{d}{dt}n_0(t) = -\lambda_{0\to\pi}\, n_0(t) + \lambda_{\pi\to 0}\, n_\pi(t), \tag{5.78}$$

$$\frac{d}{dt}n_\pi(t) = \lambda_{0\to\pi}\, n_0(t) - \lambda_{\pi\to 0}\, n_\pi(t), \tag{5.79}$$

where the subscripts on n indicate the two allowed values of θ. Solving Eqs. (5.78) and (5.79) we may derive for the time-dependent magnetization M(t):

$$M(t) = \mu V(n_0(t) - n_\pi(t))$$

$$= M(t=0)\, exp(-\bar{\lambda}\, t) + \mu V \frac{\Delta\lambda}{\bar{\lambda}}(1 - exp(-\bar{\lambda}\, t)), \tag{5.80}$$

where $(n_o + n_\pi)$ is normalized to unity,

$$\bar{\lambda} = \lambda_{0\to\pi} + \lambda_{\pi\to 0}, \tag{5.81}$$

and

$$\Delta\lambda = \lambda_{\pi\to 0} - \lambda_{0\to\pi}. \tag{5.82}$$

Here $\lambda_{0\to\pi}$ equals τ_+^{-1} whereas $\lambda_{\pi\to 0}$ equals τ_-^{-1} (cf., Eq. (5.11)). Furthermore, note that if M(t = 0) equals zero and $\mu H \ll K$ (linear response considered) Eq. (5.80) yields the earlier written expression in Eq. (5.12) wherein $M_{eq}(H,t)$is given by a combination of Eq. (5.3) and Eq. (5.7). Thus in the 'high barrier – weak noise' Kramers limit the two-state rate theory yields completely equivalent results as can be independently obtained from the Fokker-Planck description.

Interpretation of the Mössbauer Spectra

The Mössbauer data shown in Fig. 5.16 and Fig. 5.17 are interpreted on the basis of the following stochastic model Hamiltonian:

$$\mathcal{H} = -\mu_N g(I) I_z h(t), \tag{5.83}$$

where μ_N is the nuclear magneton, g(I) is the level-specific g-factor depending on whether the 57_{Fe} nucleus is in its excited state I = $\frac{3}{2}$ or the ground state I = $\frac{1}{2}$, I_z is the component of the nuclear angular momentum along the anisotropy direction, and h(t) is a local field at the nucleus that jumps about stochastically in time. The local field h(t) at the nucleus is of course produced by the magnetization of the particle which, for reasons outlined earlier, is taken to jump between $\pm h_o$, where h_o is proportional to H, given by Eq. (5.77). The rates at which these jumps occur are given by $\lambda_{0\to\pi}$ and $\lambda_{\pi\to 0}$. The Mössbauer lineshape as a function of the frequency ω is then given by [121, 122].

$$I(\omega) = \frac{1}{\pi} Re \sum_{m_0 m_1} |< I_0 m_0 | \mathcal{A} | I_1 m_1 > |^2$$

$$\int dV\, f(V) \left[\left(-i\omega + \frac{\Gamma}{2} \right) + \frac{\mu_N^2 h_0^2 (g_0 m_0 - g_1 m_1)^2}{\left(-i\omega + \frac{\Gamma}{2} \right) + \frac{\bar{\lambda}(V)}{2}} \right]^{-1} . \quad (5.84)$$

In Eq. (5.84) $\mathcal{A}$ is the nuclear transition operator, the matrix elements of which are given by Clebsch-Gordan coefficients, Γ is the natural line-width of the excited level, g's are the ground and excited level g-factors and $\bar{\lambda}(V)$ is the volume-dependent rate given by Eq. (5.81). The integral over V incorporates the particle size distribution with the aid of a probability function f(V).

As we mentioned earlier in connection with Fig. 5.15, the spectra are indicative of the presence of an apparently static magnetic field which is an internal field due to the dipolar interaction. The significant feature of the dipolar coupling is that the mean field H occurring in the exponent of the relaxation rates $\lambda_{0\to\pi}$ and $\lambda_{\pi\to 0}$ is itself volume-dependent (cf., Eq. (5.77)). Therefore, the relaxation rates acquire an additional V^2-dependent term in the exponent, for the 'Interacting' sample 1. Thus, λ for a given temperature is systematically smaller for the interacting case (sample 1) than for the noninteracting case (sample 2). Hence, even at the room temperature the sample 1 displays a split spectrum, as discussed before. We may now summarize our interpretation of the observed Mössbauer spectra as follows. At any given point of time, half of Mössbauer nuclei find their 'local' magnetic field pointing along $\theta = 0$ while the other half will see the local field pointing along $\theta = \pi$. Both these orientations would of course yield an identical six-finger pattern, because the nucleus as an observer cannot distinguish between $\theta = 0$ and

$\theta = \pi$, provided the local field is static, within the Mössbauer measurement time. That of course is determined by the temperature T and the mean dipolar strength parametrized by Λ. These two quantities determine the rates of relaxation of the dipolar field. Hence, for sample 2, for which the dipolar field is negligible, semblance of a six-finger pattern shows up only at the lowest T when relaxation slows down, whereas for sample 1, the dipolar coupling keeps the relaxation slow within the nuclear time-window, at all temperatures.

5.4. Analysis of Results

In the discussion of Mössbauer spectroscopy our focus has been the influence of dipolar interaction on slowing down relaxation. In this section we will turn our attention onto the effect of polydispersity, viz., volume-distribution of the nanoparticles. We will show that such a distribution leads to striking memory effects in low temperature DC magnetization measurements – surprisingly, and somewhat counter intuitively, dipolar interaction suppresses memory effects.

The magnetization measurements are carried out in accordance with the following cooling and heating protocol (cf. Fig. 5.18). Starting from room temperature $T = 300K$ ($T = T_{\infty}$, for our purpose), a small magnetic field of 100 Oe is applied and the magnetization (M) measured. Keeping the field on, the temperature (T) is lowered continuously at a steady rate of 2^0K/min and M is simultaneously measured upto the temperature T_n (the solid line). At T_n, the field is switched off and the drop of M is monitored for nearly 4 hours (dashed line). Subsequently, the field is switched back on and M(T) versus T is mapped in the cooling regime. Again at T_{n-1} the field is switched off and the process of measurement is repeated, until the lowest temperature T_0 is reached. Thus, one obtains field-cooled response and zero-field relaxation of the magnetization. At the end of the cooling cycle, at T_0, the field is turned on (dashed-dotted line), and M(T) is monitored as the system is heated from T_0 through T_{n-2}, T_{n-1}, T_n and eventually to T_{∞}, again at the rate of 2^0K/min, the magnetic field remaining on, throughout. Our results are shown in Fig. 5.19, for samples 1, 2 and 3 (whose attributes are

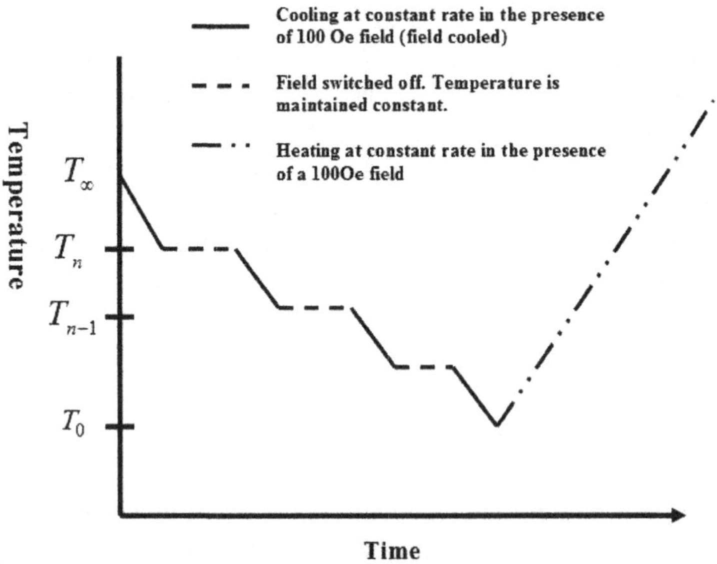

Fig. 5.18. Time-temperature protocol.

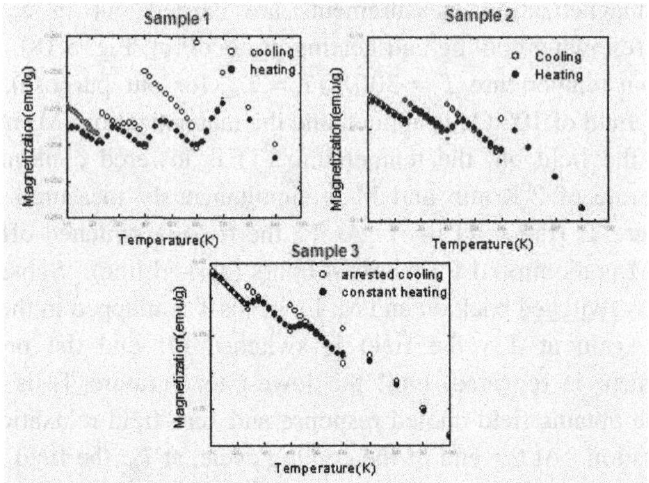

Fig. 5.19. Magnetization versus temperature data during cooling and heating cycles, for samples 1, 2 and 3.

mentioned in Fig. 14 and Table 1). The heating path surprisingly shows wiggles in M(T) at all the T steps T_{n-2}, T_{n-1}, T_n where the magnetic field was earlier switched off during cooling, apparently retaining a memory of the temperature steps at which cooling was arrested. More surprising, however, is the fact that memory effects, defined here in terms of overlap of the data during cooling and heating cycles, are the most prominent for sample 3. Recall that for the latter dipolar interaction is negligible. Indeed memory effects are the weakest for the 'Interacting' sample 1.

Our interpretation is that the observed memory effects are due to polydispersity, most prominent in sample 3, which leads to a superposition of relaxation times, because the volume V occurs in the exponent (cf. Eq. (5.10)). To illustrate this point, we now consider simulation data, for just two sizes (bidispersive), schematically depicted in Fig. 5.20. Recall that the 'Blocking Temperature' T_B is that temperature at which the experimental timescale of measurement equals the relaxation time, i.e.

$$\tau_E = \tau_0 \exp\left({}^{KV}\!/_{k_B T_B}\right), \tag{5.85}$$

if the influence of any external (or internal) magnetic field on the relaxation time is ignored. Since τ_E is fixed, the blocking temperature T_B is directly proportional to the volume V, for a given anisotropy K. Thus the smaller sized particle of volume V_1 has a lower blocking temperature than the larger sized particle of volume V_2, as in Fig. 5.20 (i.e. $T_{B1} < T_{B2}$). Now we can contrive a situation in which T_{B1} and T_{B2} lie on either side of the temperature of arrest, given by one of the dashed lines in Fig. 5.18, when the external field is switched off and the zero-field relaxation of the magnetization is monitored as a function of time. Clearly, the particle of volume V_1 will exhibit facile (superparamagnetic) relaxation whereas the particle of volume V_2 will remain relatively frozen. The observed decay of the magnetization will be a super position of these two kinds of almost extreme behavior.

We now consider magnetization as a function of the temperature T. Even at the lowest measured temperature T_0, the small particle (of volume V_1) is contrived to have its relaxation time $\tau_1$smaller that τ_E, while τ_2 for the bigger particle is larger than τ_E. Thus the smaller

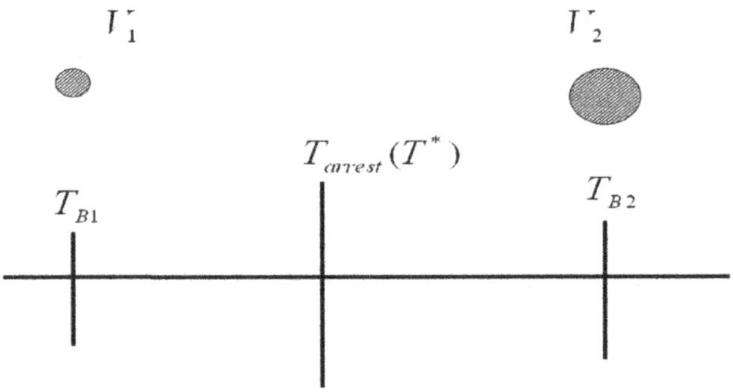

Fig. 5.20. Just two sizes corresponding to two blocking temperatures T_{B1} and T_{B2}, flanking T* on either side.

particle will equilibrate rapidly even at T_0, while the bigger particle will be 'blocked'. The crossover temperature is the temperature of arrest and when the measurement temperature is much larger than the temperature of arrest, such as at the room temperature, both τ_1 and $\tau_2$are expected to be smaller than τ_E.

With this background we turn to our M-T diagram in Fig. 5.21. Assume the temperature of arrest to be around 40K, indicated by a cross (*) in Fig. 5.20. For reasons mentioned earlier the smaller particle has its relaxation time $\tau_1$smaller than τ_2 and hence, is in thermal equilibrium with the surroundings at all measured temperatures. Therefore, the smaller particle displays Curie behavior $\left(M \propto \frac{1}{T}\right)$, both during the cooling and the heating cycles, as indicated in the right hand side of Fig. 5.21.

However, consider now more carefully the magnetization response for the larger particle. While for temperatures bigger than the temperature of arrest (T*), even the larger particle equilibrates reasonably rapidly, within the time of measurement, and hence exhibits the Curie law, this behavior below the arrest is quite different. Recall that during the cooling cycle (the upper curve on the right hand side of Fig. 5.22) the magnetic field is switched off at T*. Hence if the plateau of waiting (dashed line in Fig. 5.18) is long enough the magnetization will relax to zero. But, now when the field is restored again and the temperature is lowered the larger particle is 'blocked' off its relaxation path. Hence the

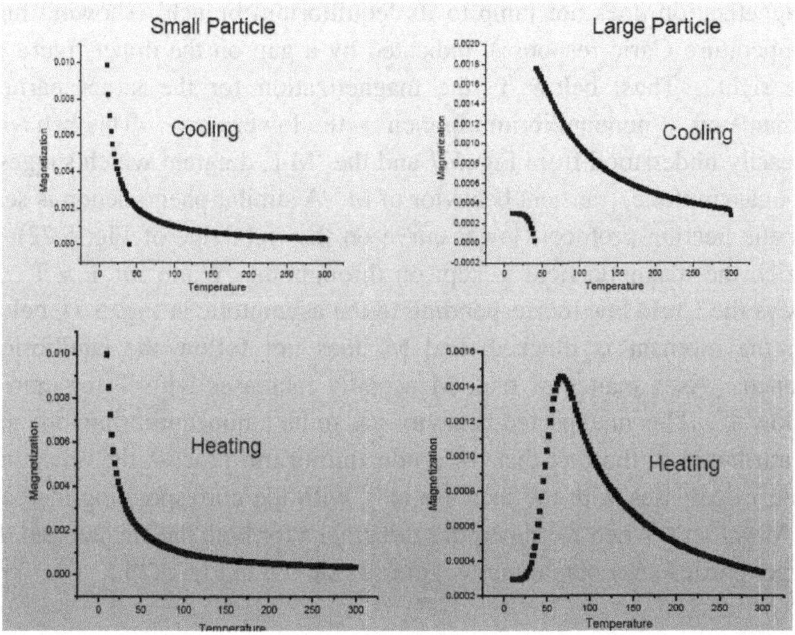

Fig. 5.21. M-T diagrams during 'cooling' and 'heating' for small and large particles.

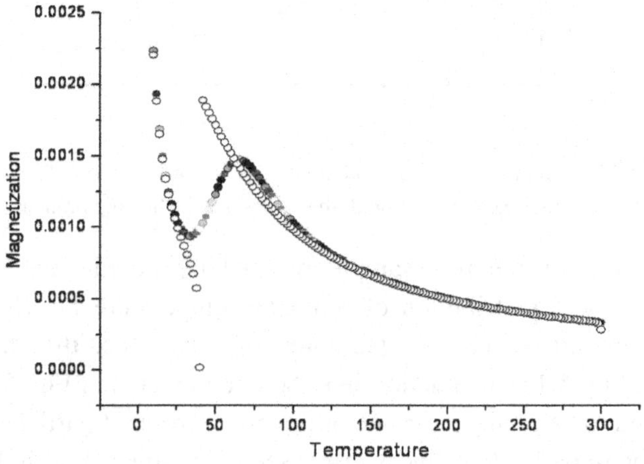

Fig. 5.22. Juxtaposition of the cooling and heating cycle results for samples 1 and 2.

magnetization does not jump to its 'equilibrium branch' (showing high temperature Curie response), indicated by a gap on the upper figure on the right. Thus, below T^* the magnetization for the larger particle remains on a 'nonequilibrium branch' – the lower curve. This behavior is easily understood from Fig. 5.7 and the 'M-t' diagram which suggests a nonasymptotic, transient behavior of M. A similar phenomenon is seen for the heating protocol (lower curve on the right side of Fig. 5.22), in which the magnetic field is kept on throughout. While for $T > T^*$, M obeys the Curie law (corresponding to the asymptotic in Fig. 5.7), below T^*, the moment is blocked, and M does not follow the equilibrium branch. As a matter of fact, M actually increases with T for upto T below T^*. This unexpected behavior has to be a nonequilibrium one and is attributed to the fact that for a nonequilibrium process the relaxation time τ decreases with the increase of T with the corresponding increase of M-values. When we superpose the graphs for both the smaller and the larger particles we obtain the 'wiggle', as shown in Fig. 5.23.

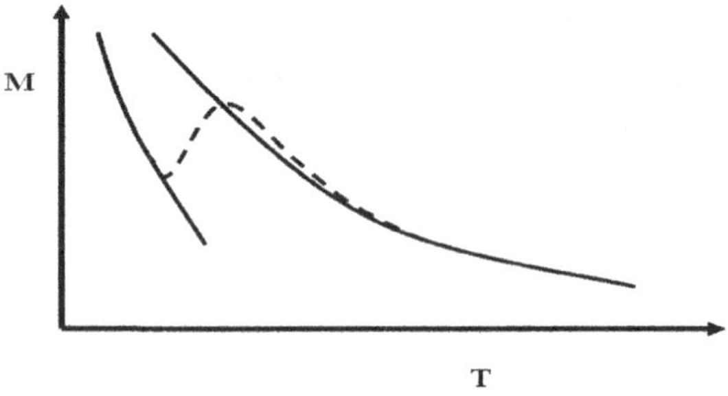

Fig. 5.23. M-T diagram for the small and large particles of Fig. 5.20. The solid lines are the cooling branch while the dashed lines are the heating branch.

Evidently, for the real sample marked by polydispersity there is expected to be a distribution of blocking temperature T_B. Hence, even though there are several temperatures of arrest (i.e. the flat, dashed regions in Fig. 5.18), a situation like the one depicted in Fig. 5.23 would be expected to be realized for one or the other group of particles. The net result is the remarkable memory effect seen for sample 3, as in Fig. 5.19.

Why is then the memory effect for sample 2 not as spectacular as for sample 3, even though both the samples have their average inter-particle distance $\simeq$15nm, making them 'noninteracting'? Our interpretation is that the sample 3 has an average particle size 5 times that of sample 2 with corresponding higher polydispersity. As we argued in the preceding paragraphs, polydispersity is crucial for the overlap of the cooling and heating data, and hence memory, for all values of the temperatures of arrest.

So far we have not brought in the issue of dipolar interaction, thought to be absent in samples 2 and 3. But, what about sample 1 in which the interaction effects are expected to be more pronounced, as verified by the Mössbauer spectra discussed earlier? Well, the memory effects, defined in the sense discussed above, seem blurred, for sample 1. The reason? We have discussed earlier at the end of the last section that the dipolar interaction contributes a V^2- term in the exponent of the relaxation time, thus slowing down relaxation. This effect, coupled with diminished polydispersity for sample 1, make the situation depicted in Fig. 5.20 not as probable as for sample 3. Therefore, for the kind of memory effects we have seen it is crucial to have combination of two groups of particles – one group that equilibrates rapidly (and hence shows superaramagnetism) at all temperatures of interest and the other group which has mixed superparamagnetic and frozen relaxation behavior, depending on the temperature.

In summary, we have discussed in this section the symbiotic relationship of polydispersity and interaction in influencing relaxation phenomena in single-domain nanomagnetic particles. We have demonstrated that just a bimodal distribution of particle size, in which one set of particles remains frozen in its response behavior while the other set exhibits magnetic viscosity, suffices to interpret dramatic memory effects seen in cooling and heating cycles of the magnetic response. Our results further suggest that either by tuning the interaction (through changing inter-particle distance) or by tailoring the particle size distribution, these nanosized magnetic systems can be put to important applications in memory devices. In the spirit of these Lecture Notes we should also mention that nanomagnetic particles constitute yet another magnetic paradigm for studying and understanding similar memory phenomena in very different systems of Relaxor Ferroelectrics and Shape Memory Ferroelastics.

Chapter 6

Dissipative Quantum Systems

6.1. Introduction

A pure quantum system is associated with a unitary evolution in time of its wave function because the underlying Hamiltonian is Hermitean. This characteristic unitarity is indeed at the heart of important phenomena such as interference and coherence which arise from the fact that the phase of the wave function has definite relation at two different times. The coherence effects are responsible for such fascinating properties as persistence currents in mesoscopic rings, Bohm – Aharonov phase and quantum Hall effect in the condensed matter, just to name a few [123, 124]. For most real systems however dissipation sets in because the system under focus cannot be decoupled from its environment or other degrees of freedom which are not the objects of direct study. The reason that quantum coherence effects can be observed in mesoscopic and nanoscopic systems is because the dimension of these systems is smaller than the ('inelastic') mean free path, which measures the mean distance an electron has to traverse before it suffers an energy-losing collision. But even without violation of this condition a quantum system can lose its coherence because of say the finiteness of the temperature of the system. At a finite temperature thermal vibrations or phonons are excited which, through interaction with say the itinerant electrons in a metallic wire, can introduce decoherence. The point is, when one is focusing on the transport properties of the electrons, one is not at the same time looking at the phonons and therefore when the phonons are averaged out from a larger Hilbert space the projected dynamics of the electrons belonging to a truncated Hilbert space do appear to be dissipative, because of decoherence effects.

The problem we are trying to address therefore belongs to the larger issue of how a quantum system transits from one equilibrium state to another because certain conditions, such as the temperature, are altered. This question is indeed central to the subject of nonequilibrium statistical mechanics of a quantum system. The latter is normally studied in the so-called System-plus-Bath approach in which the system under focus is viewed to be embedded in a bath with which it is in strong contact [87]. While it is the set of system degrees of freedom which is normally under the focus of study the bath degrees of freedom do not come under direct reckoning and are averaged over. This averaging process however leads to dissipative phenomena. The issue at hand is akin to the classical Brownian motion of a pollen particle wading its way through a fluid with which it suffers collisions [111]. The pollen particle is like the 'quantum' electron or a spin whereas the fluid degrees of freedom are similar to phonons. In that sense the Brownian motion of a pollen particle is entirely analogous to the ohmic conduction of an electron, the viscosity of the Brownian fluid playing a similar role as the resistance of a metal. In the quantum problem, therefore one has to deal with "Quantum Brownian Motion", tackled through, say, a quantum Langevin equation [125].

In the context of the System-plus-Bath approach two distinct cases merit separate attention. In the first, the bath can be approximated as classical while in the second the quantum nature of the bath enters in an essential way, though in both cases the system under focus is quantum mechanical. Considering the theme of these Lecture Notes it is not surprising that the first example of a quantum dissipative system that comes to our mind belongs to Magnetism, in the context of spin-lattice relaxations [126, 127]. As we will see later, a spin is merely a metaphor – it may also mean a two – or a three-level atom in Quantum Optics, and relaxation or damping may be occasioned not by phonons but electromagnetic radiation fields [90]. Thus we refer back to Chapter 4 and revisit the problem of the relaxation of a single spin, dealt with earlier via stochastic methods, but now subjected to a full quantum mechanical treatment. We may recall from Chapter 4 that relaxation of a spin was studied in terms of a probability function, which was entirely classical. For a quantum system, however, the phase is an extremely crucial property as alluded to earlier and hence, the quantity that replaces the probability is the density operator, the off-diagonal elements of which are a measure of the quantum phase. Our starting point then, in Sec. 6.2, is the Liouville Von Neumann equation for the density operator. We show that when the bath is averaged over we arrive at Bloch equations for the longitudinal and transverse components of the magnetization. Though these equations were first discussed in the context of Nuclear Magnetic Resonance, they have wider applicability to Quantum Optics – the 'up' and 'down' states of a single spin are like the two energy states of an atom, for instance.

In the usual derivations of the Bloch equations, as summarized in Sec. 6.2, the bath is treated classically. But, as we had mentioned before, important phenomena emerge when the bath itself has quantum attributes. For instance, consider the bath to comprise a bunch of quantum harmonic oscillators. Because a quantum oscillator has 'zero point motion' the dynamics of the bath do not stop, even at zero temperature! This property is responsible for say a quantum particle, normally tunneling between two wells of a double-well, getting trapped in one of the wells when the coupling with the bath exceeds a critical value [128]. Therefore, the study of quantum relaxation in which the bath has essential quantum properties has given rise to a new field of activity called Dissipative Quantum Mechanics [129]. The latter subject is of great relevance in the investigation of dissipative tunneling, mentioned above, as well as macroscopic quantum coherence [130] and indeed the topically important issue of quantum information processes [131]. Once again we will see that the consequent generalization of single spin relaxation yields the so-called 'Spin-Boson Hamiltonian' that has formed a standard paradigm for dissipative quantum mechanics [132]. This is the subject of Sec. 6.3.

We referred earlier to the quantum Brownian motion and the concomitant quantum Langevin equation. The accompanying issues are best exemplified by the quantum response of the orbital motion of an electron to an external magnetic field and how the resultant diamagnetic moment gets dissipated because of the 'Brownian motion' of the orbiting electron, occasioned by the contact with a quantum heat bath. The dissipative Landau diamagnetism, the topic of Sec. 6.4, is therefore another important paradigm for studying dissipative quantum mechanics in general and the coherence – decoherence phenomenon in particular [133].

The usual scenario for the transition from coherence to decoherence or from quantal to classical behavior is via the increase of temperature or the coupling with the environment. However, even when the latter two parameters are held fixed, one can see quantum – classical crossover by enlarging the value of the spin magnetic moment. The point is, a spin one-half entity is most quantal and a spin infinity system is most classical; where is the cross-over? The issue is best studied in molecular or cluster magnets which are the miniature version of the nano-magnetic particles considered in Chapter 5. The effective spin in these systems can be as large as say S = 10 and because the spin is composed of several atomic spins, the associated magnetization may be viewed 'mesoscopic' [134]. We are thus led to study mesoscopic tunneling of the magnetization as the spin angular momentum makes transitions amongst, say the 21 states of the S = 10 system, induced by a quantum heat bath [135]. This is the topic of our conducting Sec. 6.5.

6.2. Nonequilibrium Statistical Mechanics

The standard prescription in the System-plus-Bath approach to statistical mechanics is to assume that initially, at time t = 0, the system under study is characterized by an arbitrary density operator denoted by $\rho_S(o)$ and is decoupled from its environment, viewed as a heat bath. The latter is described by a density operator appropriate to the Gibbsian canonical ensemble of statistical mechanics for a system in thermal equilibrium at a temperature T. The system plus the bath are thereby assumed to belong to a product Hilbert space and therefore, the composite density operator is given by

$$\rho(t = 0) \equiv \rho(o) = \rho_S(o) \otimes \frac{e^{-\beta \mathcal{H}_B}}{Z_B}, \tag{6.1}$$

where $\mathcal{H}_B$ is the Hamiltonian for the bath, assumed to be at a fixed temperature T ($\beta \equiv (K_B T)^{-1}$) and Z_B is the corresponding partition function given by

$$Z_B = Tr\left(e^{-\beta \mathcal{H}_B}\right). \tag{6.2}$$

The subsequent time evolution of the overall density operator is given as usual by the Liouville – Von Neumann equation:

$$i\hbar \frac{\partial}{\partial t}\rho(t) = [\mathcal{H}, \rho(t)], \tag{6.3}$$

where $\mathcal{H}$ is the total Hamiltonian:

$$\mathcal{H} = \mathcal{H}_s + V + \mathcal{H}_B. \tag{6.4}$$

The underlying philosophy is therefore to imagine that the interaction V between the system and the bath is switched on at time $t = o^+$, and the theory is contrived such that asymptotically as t approaches infinity, the system comes to thermal equilibrium with the bath. As discussed below, this envisaged scenario is expected to be enacted by assuming the coupling term V to be weak, treatable in perturbation theory. The weakness of the coupling is taken to imply that the equilibrium of the bath is not at all disturbed, which presupposes that the bath comprises an infinitely large number of degrees of freedom that can absorb, through infinite channels, the energy infused by the system without undergoing any change in temperature.

In the magnetic paradigm of a single spin, discussed in Chapter 4, $\mathcal{H}_s$ is given by

$$\mathcal{H}_s = -\mu \mathrm{H} \mathrm{S}_z, \tag{6.5}$$

where H is the external magnetic field applied along the z-axis in the laboratory. While the bath was treated stochastically earlier, now its dynamics is described explicitly by a Hamiltonian $\mathcal{H}_B$. However, in tune with the assumed picture that the relaxation of the system proceeds by single spin flips, induced by the bath, the coupling term V can be taken to be of the form:

$$\mathrm{V} = \mathrm{g}\hat{\mathrm{b}}\mathrm{S}_x, \tag{6.6}$$

where g is a coupling constant and $\hat{b}$ is an operator which acts on the Hilbert space of the bath. Because z has been chosen as the axis of quantization, S_x is purely off-diagonal in the representation in which S_z is diagonal and hence, the chosen form of the interaction in Eq. (6.6) can indeed guarantee spin-flip processes.

Coming back to the general problem posed by Eqs. (6.3) and (6.4) and introducing interaction-picture operators as

$$\rho_I(t) = e^{\frac{i}{\hbar}(\mathcal{H}_s+\mathcal{H}_B)t}\rho(t)e^{-\frac{i}{\hbar}(\mathcal{H}_s+\mathcal{H}_B)t}, \tag{6.7}$$

$$V_I(t) = e^{\frac{i}{\hbar}(\mathcal{H}_s+\mathcal{H}_B)t}\, Ve^{-\frac{i}{\hbar}(\mathcal{H}_s+\mathcal{H}_B)t}, \tag{6.8}$$

Eq. (6.3) may be rewritten as

$$i\hbar\frac{\partial}{\partial t}\rho_I(t) = [V_I(t), \rho_I(t)]. \tag{6.9}$$

At this stage it is convenient to introduce the Liouville operator L_I appropriate for V_I and rewrite Eq. (6.9) as

$$\frac{\partial}{\partial t}\rho_I(t) = -iL_I(t)\,\rho_I(t), \tag{6.10}$$

with the formal solution:

$$\rho_I(t) = \exp_T\left(-i\int_0^t L_I\,(t')dt'\right)\rho(o). \tag{6.11}$$

In Eq. (6.11), exp_T denotes a time-ordered series where the operators are placed from the left to right in decreasing order of their time-arguments. Equation (6.7) then yields

$$\rho(t) = e^{-i(L_s+L_B)t}exp_T(-i\int_0^t L_I(t')dt')\,\rho(o), \tag{6.12}$$

where L_s and L_B are the Liouville operators associated with $\mathcal{H}_s$ and $\mathcal{H}_B$ respectively.

We are now ready to carry out an important step that amounts to extracting from Eq. (6.12) a 'reduced' density operator for the system alone by averaging over or projecting out the bath degrees of freedom:

$$\rho_S(t) = Tr_B[\rho(t)], \tag{6.13}$$

where "Tr_B" denotes the trace operation over the bath variables. Equation (6.12), in conjunction with the factorized density operator in Eq. (6.1), then yields

$$\rho_S(t) = e^{-iL_st}[exp_T(-i\int_0^t L_I(t')dt')\,]_{av}\rho_S(o), \tag{6.14}$$

where the term e^{-iL_Bt} drops out from further reckoning because of the cyclic property of the trace operation over the Hilbert space of the bath. In terms of explicit matrix elements,

$$[exp_T(-i\int_0^t L_I(t')dt')]_{av}$$

$$= \sum_{n_b n_b'} (n_b n_b| \, exp_T(-i\int_0^t L_I(t')dt')|n_b' n_b') < n'_b|\frac{e^{-\beta\mathcal{H}_B}}{Z_B}|n'_b>, \tag{6.15}$$

where the states $|n_b>, |n_b'>$ refer to the eigenstates of the bath Hamiltonian $\mathcal{H}_B$.

In accordance with our stated objective as spelt out in the introductory remarks, we want to treat the interaction term V perturbatively. To this end we apply the cumulant-expansion theorem [91] on Eq. (6.15) and retain terms upto second-order cumulants only. Thus

$$\rho_S(t) = e^{-iL_st}\{exp_T(-i\int_0^t dt' < L_I(t') >_{av} - \int_0^t dt'\int_0^t dt''$$
$$[< L_I(t')L_I(t'') >_{av} - < L_I(t') >_{av} < L_I(t'') >_{av}])\}\rho_s(o). \tag{6.16}$$

In the above, $< \cdots >_{av}$ denotes an averaging over the bath degrees of freedom, as in Eq. (6. 15). In addition, the subscript T implies that the time-ordering has to be maintained in writing the second-order cumulant. Thus, for instance,

$$\langle L_I(t')L_I(t'')\rangle_{av} = \sum_{n_b,n_b'} (n_b n_b|L_I(t')L_I(t'')|n_b' n_b')\,\langle n_b'|\frac{e^{-\beta\mathcal{H}_B}}{Z_B}|n_b'\rangle. \tag{6.17}$$

Without any loss of generality we can take $< L_I(t') >_{av}= 0$ by assuming $< V_I >_{av}= 0$. Even if the latter were finite, $< L_I(t') >_{av}$, being time-independent (i.e. independent of t'), can be easily absorbed in the unperturbed term. Thus, we take the system to be invariant under time translation, i.e. we assume the system to be stationary, and further obtain from Eq. (6.16),

$$\rho_s(t) = e^{-iL_st}\exp\,(-\int_0^t d\tau(t-\tau) < L_I(\tau)L_I(o) >_{av})\rho_s(o). \tag{6.18}$$

The second-order perturbation theory, as carried out above, plus the assumption of stationarity, goes under the name of the Born-Markov approximation in the context of Quantum Optics [90]. Indeed the time-derivative of Eq. (6.18) can be cast into the form of a much-studied master equation [88].

The correlation function in Eq. (6.18) contains the effect of heat-bath induced fluctuations. If the latter are assumed to be operative on time-scales much shorter than the time-scales of interest for the system dynamics, the upper limit of the integral in Eq. (6.18) can be extended to infinity. The result is

$$\rho_s(t) = e^{-iL_st}\exp\left(-t\int_o^\infty d\tau < L_I(\tau)\,L_I(o) >_{av}\right)\rho_s(o), \tag{6.19}$$

and hence,

$$\frac{\partial}{\partial t}\rho_s(t) = -\frac{i}{\hbar}[\mathcal{H}_s, \rho_s(t)] - e^{-\mathcal{H}_s t/\hbar}\,\hat{R}\,e^{-\mathcal{H}_s t/\hbar}\rho_s(t), \tag{6.20}$$

where $\hat{R}$ is the so-called quantum relaxation matrix:

$$\hat{R} = \int_0^\infty d\tau < L_I(\tau)\, L_I(0) >_{av}. \tag{6.21}$$

The literature abounds with different treatments of Eq. (6.16) leading to various forms of the master equation [88].

(i) Bloch Equations:

It is instructive to illustrate the application of the master equation by considering the case of single-spin relaxation for which the unperturbed Hamiltonian and the interaction terms are given by Eqs. (6.5) and (6.6) respectively. The reduced density operator is now a 2 x 2 matrix, with both diagonal and off-diagonal elements. These can be employed to derive equations of motion for the magnetization. Thus for instance,

$$\frac{d}{dt}m_z(t) = \mu\frac{d}{dt}Tr(\rho_s(t)s_z) = \mu\frac{d}{dt}[< +|\rho_s(t)|+> - < -|\rho_s(t)|->]. \tag{6.22}$$

The right hand side of Eq. (6.22) can be evaluated from the master equation (6.20). After some algebra, we can derive [88]

$$\frac{d}{dt}m_z(t) = -\lambda[m_z(t) - \mu\tanh(\beta\mu H)]. \tag{6.23}$$

While Eq. (6.23) is identical to Eq. (4.35), derived earlier from stochastic considerations in which λ is interpreted as the relaxation rate, now λ can be derived from microscopic considerations. Thus we have [88]

$$\lambda = 2\frac{g^2}{\hbar^2}\int_0^\infty d\tau \cos(2\mu H\, {}^{\tau}/_{\hbar}) < \hat{b}(\tau)\hat{b}(0) >, \tag{6.24}$$

where the time-dependent bath operators have their usual Heisenberg picture representations:

$$\hat{b}(\tau) = e^{\frac{i}{\hbar}\mathcal{H}_B\tau}\hat{b}(o)e^{-\frac{i}{\hbar}\mathcal{H}_B\tau}. \tag{6.25}$$

Therefore, the relaxation rate λ, which appeared earlier as a phenomenological parameter, has the form of a power spectrum of heat bath fluctuations.

While the stochastic methods are incapable of dealing with the transverse components of the magnetization as the latter are associated with the off-diagonal terms of the density operator, we now have,

$$\frac{d}{dt}m_{x,y}(t) = \mu\, Tr\left(\frac{\partial \rho_s(t)}{\partial t} S_{x,y}\right). \tag{6.26}$$

From the master equation (6.20) we can derive [88]

$$\frac{d}{dt}m_x(t) = \frac{2H\mu}{\hbar}m_y(t) - \lambda m_x(t),$$

$$\frac{d}{dt}m_y(t) = -\frac{2H\mu}{\hbar}m_x(t) - \lambda m_y(t). \tag{6.27}$$

Note that while the longitudinal component of the magnetization has relaxational dynamics only, the transverse components are endowed with oscillatory dynamics as well, on a frequency scale $2H\,{}^{\mu}/_{\hbar}$.

Equations (6.23) and (6.27) are called the Bloch equations that have wide applications to the dynamics of a two-level atom, in the context of Quantum Optics. The oscillations accompanying the transverse components are known as the Rabi oscillations [90]. It is pertinent to point out that the operator $\hat{b}$ pertaining to the bath, for a two-level atom, refers to the creation or annihilation operator for the electro-magnetic fields surrounding the atom. Recall that it is the interaction with these electromagnetic fields that is responsible for the spontaneous decay of an atom from the excited to the ground state.

6.3. Spin-Boson Hamiltonian

In the above treatment of the dynamics of a single quantum spin the heat bath Hamiltonian was not discussed in detail. As mentioned in the introduction (Sec. 6.1) recent years have witnessed a spurt of activity in the context of dissipative quantum systems for which the bath Hamiltonian has to be treated explicitly and quantum mechanically. Once again the standard paradigm can be found in magnetism in framing the simple question: how does a spin-half system get magnetized when an external magnetic field is applied? The magnetic field causes a Zeeman splitting and at finite temperatures the lower of the two levels is preferentially populated in accordance with Boltzman factors. Thus there is a redistribution of population which must involve quantum transitions from the higher to the lower level. These transitions are inevitably accompanied by simultaneous transitions or redistribution of energies in the environment of the spin system. The most common form of the environment is one of thermal vibrations whose quanta are phonons and the relevant process is called the spin-lattice relaxation. We may again invoke the system-plus-bath approach in which the total Hamiltonian can be written as:

$$\mathcal{H} = \mathcal{H}_s + V + \mathcal{H}_B, \tag{6.28}$$

$$\mathcal{H}_s = -\mu H S_z, \tag{6.29}$$

$$V = S_x \sum_{k} g_{\underline{k}} \left(b_{\underline{k}} + b_{\underline{k}}^{+} \right), \tag{6.30}$$

$$\mathcal{H}_B = \sum_{\underline{k}} \hbar\omega_{\underline{k}}\, b_{\underline{k}}^{+} b_{\underline{k}}. \tag{6.31}$$

The term V, written in second-quantized notation, describes spin-phonon interactions via a coupling constant $g_{\underline{k}}$ for the $\underline{k}^{th}$ phonon mode. The Hamiltonian $\mathcal{H}_B$ is that of a phonon bath with a prescribed phonon dispersion governed by the $\underline{k}$-dependence of the frequency $\omega_{\underline{k}}$. The interaction term, being proportional to S_x, is off-diagonal in the representation for which z, along which the magnetic field H is applied, is the quantization axis. This term then causes quantum transitions between the energy levels of $\mathcal{H}_s$. The strength of these transitions is proportional to a phonon displacement field, which in turn is modulated by the bath Hamiltonian $\mathcal{H}_B$. The study of spin-lattice relaxation is of great importance in the context of nuclear magnetic resonance and electron paramagnetic resonance [136].

It is interesting to note that while the problem posed above is set in the context of magnetism a rotation in spin angular momentum space by an angle $\pi/2$ about the Y-axis turns the problem into an equivalent one which is appropriate for studying the effect of dissipation on the tunneling states of a particle moving in a double-well. The equivalent problem is defined by the Hamiltonian:

$$\mathcal{H} = \Delta S_x + S_z \sum_k g_{\underline{k}} \left(b_{\underline{k}} + b_{\underline{k}}^{+}\right) + \sum_k \hbar\omega_{\underline{k}} b_{\underline{k}}^{+} b_{\underline{k}}. \tag{6.32}$$

This Hamiltonian describes a hydrogen atom, proton or a positive muon that is trapped amongst two equivalent interstitial sites in a solid [137]. These two sites can be viewed as the two minima of a double well, shown schematically in Fig. 6.1. The interstitial atom can tunnel between the two sites, even at zero temperature. If the two minima are represented by two eigenetates of a pseudo spin operator S_z then the tunneling Hamiltonian can be represented by the first term of Eq. 6.32, Δ being the tunneling frequency. The interstitial particle distorts the site it occupies and carries the distortion field, described by phonons, as it tunnels, much like a hopping polaron [138]. Clearly, the nature of the phonons involved in the process, acoustic or optic, will be dictated by the chosen form of

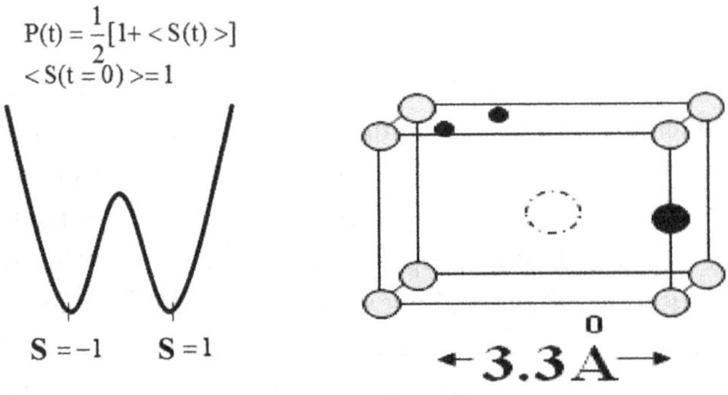

Fig. 6.1. A model for two possible sites for H (smaller black dots) in the bcc structure of Nb (corner atoms) in the presence of trapping atoms e.g. O (larger black dot). The H tunnels in a double well, shown at the left of Fig. 6.1.

the spectral density of phonons. In a different context, the Hamiltonian in Eq. (6.32) may also describe tunneling centers in metals at low temperatures. Now the distortion fields are not phonons but are caused by conduction electrons that interact with the interstitial atom via coulomb electrons. If we account for electron-hole excitations near the Fermi surface of a metal in terms of bosons, the Hamiltonian in Eq. (6.32) is applicable again, albeit with a different spectral density for bosonic excitations [139]. Appropriately, the Hamiltonian in Eq. (6.32) has been referred to by Leggett et al as the Spin-Boson Hamiltonian [132].

A physical measure of the tunneling phenomenon is the stay-put probability $P(t)$ that a quantum particle, moving in the double-well of Fig. 6.1, will continue to stay at the right well say, if it started from that well at time $t = 0$. This quantity is defined by

$$P(t) = \frac{1}{2}[1 + < S_z(t) >], \tag{6.33}$$

where we have associated the state $|+>$ (with the eigevalue $+1$ for S_z) with the right well, and correspondingly the state $|->$ with the left well. Note that, when the tunneling particle, viewed as the 'system' in the present instance, is decoupled from its environment, i.e. the dynamics is described by the first term in (6.32) alone, $P(t)$ is given by

$$P(t) = \cos(\Delta t). \tag{6.34}$$

The tunneling particle then is like a quantum clock with frequency Δ, as it makes transitions between the two wells. This process is a quantum coherent one – the effect of the environment is to cause decoherence. The Spin-Boson Hamiltonian has been extensively studied by Leggett et al and has been reviewed by Weiss [129]. They find that the averaging process over the heat bath makes the system appear dissipative and consequently the probability function $P(t)$ is transformed from an oscillatory to an exponential function in time. Concomitantly, the tunneling behavior changes from coherent to incoherent, as mentioned above. An extreme and intriguing form of incoherent tunneling occurs when the particle does not tunnel at all, i.e. it is localized in one of the wells yielding a broken-symmetry transition. The latter is similar to the watched-put effect [140], where the two minima of the double well represent the boiling and non-boiling states of the pot – localization implies the pot does not boil at all if it is continually watched!

(i) Dilute Bounce Gas Approximation (DBGA)

As mentioned above the Spin-Boson Hamiltonian has been studied in detail by Leggett et al using functional-integral methods of quantum mechanics [132]. A major advantage of this approach is that it is non-perturbative. In that sense, Leggett et al have provided a new direction to quantum dissipative phenomena – distinct from the quantum relaxation theory, discussed in Sec. 6.2, wherein the coupling between the system and the bath has been assumed weak. However, we indicate below how the master equation derived above for spin relaxation can be suitably adapted to handle strong coupling cases as well.

The first step in the study of strong-coupling cases is the unitary transformation, well known in polaron physics [138], defined by the operator

$$\mathcal{S} = \exp\left[-\sum_{\underline{q}} \frac{g_{\underline{q}}}{2\hbar\omega_{\underline{q}}} (b_{\underline{q}} - b_{\underline{q}}^{+}) S_z\right]. \tag{6.35}$$

This transforms the Hamiltonian $\mathcal{H}$ to $\hat{\mathcal{H}}$, where

$$\hat{\mathcal{H}} = SHS^{-1} = \frac{1}{2}\Delta(S^{+}B_{-} + S^{-}B_{+}) + \sum_{\underline{k}} \hbar\omega_{\underline{k}} b_{\underline{k}}^{+} b_{\underline{k}}, \tag{6.36}$$

with

$$B_{\pm} = exp\left[\pm\sum_{\underline{k}} \frac{g_{\underline{k}}}{2\hbar\omega_{\underline{q}}} (b_{\underline{q}} - b_{\underline{q}}^{+})\right], \tag{6.37}$$

and $S^{\pm}$ are the usual ladder operators for a spin-half system.

In line with the system-plus-bath approach we can write, upon comparison of $\hat{\mathcal{H}}$ with Eq. (6.4),

$$\mathcal{H}_s = 0,$$
$$V = \frac{1}{2}\Delta(S^+B_- + S^-B_+),$$
$$\mathcal{H}_B = \sum_{\underline{k}} \hbar\omega_{\underline{k}} b_{\underline{k}}^+ b_{\underline{k}}. \quad (6.38)$$

We may recall that in the derivation of the master equation (6.20) the coupling term V is treated in the Born-Markov approximation. But since the coupling constant $g_{\underline{q}}$ appears in the exponent of the operators $B_{\underline{+}}$ a similar perturbative treatment of V will be tantamount to retaining certain types of terms, to all orders in the coupling constant, and hence, strong coupling can be accounted for. Using Eq. (6.20) we may derive [141].

$$\frac{d}{dt} < S_z(t) >= Tr\left(\frac{\partial \rho_s(t)}{\partial t} S_z\right),$$
$$= -\Delta^2 \int_0^t d\tau[\Phi(\tau) + \Phi(-\tau)]\langle S_z(t-\tau)\rangle \quad (6.39)$$

where $\Phi(\tau)$ is a correlation function involving bath operators

$$\Phi(\tau) =< B_-(o)B_+(\tau) >_{av} = < B_+(o)B_-(\tau) >_{av}. \quad (6.40)$$

The solution of Eq. (6.39) is equivalent to what is called the dilute bounce gas approximation (DBGA) within the context of the functional integral approach. A bounce is like a sojourn of the particle from say, the right well to the left well and back. If that takes a long time ($\sim\Delta^{-1}$), bounces are rare or dilute and hence it is not surprising that the derived solution in Eq. (6.39) is valid to the order Δ^2, i.e. the tunneling frequency is tacitly assumed small in comparison to bath-induced relaxation rates.

In spin-relaxation theory the bath correlation function is not calculated explicitly but is subsumed in a lumped parameter called the relaxation rate, as in Eq. (6.24). However now, with the structure of the bath Hamiltonian spelt out, it is instructive to unravel the content of the correlation function $\Phi(\tau)$. For this we need to specify the form of the frequency-dependence of the spectral density, defined by

$$J(w) = 2\sum_{\underline{q}} g_{\underline{q}}^2\ \delta\left(\omega - \omega_{\underline{q}}\right). \tag{6.41}$$

In general, $J(\omega)$ can be written as

$$J(\omega) = K\omega^s\ e^{-\omega/\omega_c}, \tag{6.42}$$

where K is a damping parameter, ω_c is a cut-off frequency for bath excitations and the exponent s characterizes the spectral nature of the bath modes. One of the most well-studied cases is that of the Ohmic dissipation model, for which $s = 1$.

For this, we have

$$\Phi(\tau) = \exp\left\{-\Sigma_{\underline{q}} \frac{4g_{\underline{q}}^2}{\hbar^2\omega_{\underline{q}}^2}\left[\coth\left(\frac{\hbar\beta\omega_{\underline{q}}}{2}\right)\left[1 - \cos\left(\omega_{\underline{q}}\tau\right)\right] + i sin\left(\omega_{\underline{q}}\tau\right)\right]\right\}. \tag{6.43}$$

The summation over $\underline{q}$ can be converted into an integral over a continuum of modes using Eq. (6.41), thus yielding

$$\Phi(\tau) = \exp\left\{-2\int_0^\infty d\omega \frac{J(\omega)}{\omega^2}\left[\coth\left(\frac{\hbar\beta\omega}{2}\right)[1 - \cos(\omega\tau)] + i\sin(\omega\tau)\right]\right\}. \tag{6.46}$$

Using the Ohmic form of the spectral density in Eq. (6.42) (with $s = 1$) we may derive

$$\Phi(\tau) = \exp\left\{-2K\int_0^\infty \frac{d\omega}{\omega} e^{-\omega/\omega_c}\left[\coth\left(\frac{\hbar\beta\omega}{2}\right)(1 - \cos\omega\tau) + i\sin\omega\tau\right]\right\}. \tag{6.45}$$

Furthermore, for $\beta\omega_c \gg |$, we have the closed-form result

$$\Phi(\tau) = \exp\ [\pi K\ sgn(\tau)]\left[\frac{\pi}{\hbar\beta\omega_c \sinh\left(\frac{\pi|\tau|}{\hbar\beta}\right)}\right]. \tag{6.46}$$

This result can be substituted for numerically solving Eq. (6.39) and the ensuing solution can be employed for the evaluation of $P(t)$, which will be discussed below.

(ii) Beyond the DBGA

The DBGA, besides being perturbative in the tunneling frequency Δ, suffers from the serious drawback that it misses the correct equilibrium behavior in the asymptotic $t \to \infty$ limit. The point is, in the system-plus-bath approach, the built-in assumptions are such that asymptotically the system described by $\mathcal{H}_S$ is equilibrated at the temperature of the bath, governed by the Hamiltonian $\mathcal{H}_B$. But, because the unitary transformation (cf., Eq. (6.35) underlying the DBGA nullifies $\mathcal{H}_S$, the equilibrium expressions remain devoid of any reference to $\mathcal{H}_S$. A simple remedy, which amounts to invoking inter-bounce terms [142], is to add and subtract the free tunneling term in Eq. (6.32) yielding [143].

$$\mathcal{H} = \frac{\Delta}{2}(S^+S^-) + \frac{1}{2}\Delta[S^+(B_- - 1) + S^-(B_+ - 1)] + \sum_{\underline{k}} \hbar\omega_{\underline{k}} b_{\underline{k}}^+ b_{\underline{k}}. \tag{6.47}$$

The basic idea behind the above decomposition is that, in any perturbative treatment of the second term in Eq. (6.47), free tunneling given by the first term, is dealt with exactly. Thus the unperturbed Hamiltonian is now given by

$$\mathcal{H}_s = \frac{\Delta}{2}(S^+ + S^-). \tag{6.48}$$

In this case, the master equation (6.20) leads to

$$\frac{d}{dt} < S_z(t) >= -\int_0^t d\tau\, [K_{11}(\tau) < S_z(t-\tau) > +K_{13}(\tau) < S^+(t-\tau) - S^-(t-\tau) >], \tag{6.49}$$

$$\frac{d}{dt} < S^+(t) - S^-(t) = -\int_0^t d\tau [4K_{13}(\tau) < S_z(t-\tau) > +K_{33}(\tau) < S^+(t-\tau) - S^-(t-\tau) >], \tag{6.50}$$

where

$$K_{11}(t) = \frac{\Delta^2}{2}[1 + \cos(2\Delta t)][\Phi(-t) + \Phi(t)],$$

$$K_{13}(t) = i\,\frac{\Delta^2}{2}\sin(2\Delta t)\,[\Phi(-t) + \Phi(t)],$$

$$K_{33}(t) = \Delta^2 \cos(2\Delta t)\,[\Phi(-t) + \Phi(t)]. \tag{6.51}$$

Note that the earlier DBGA result of (6.39) is recovered upon setting $cos(2\Delta t) = 1$ and $\sin(2\Delta t) = 0$, which again underscores our earlier assertion that DBGA is valid over time scales shorter than the inverse tunneling frequency Δ, which is another way of saying that Δ is relatively small.

The results in the DBGA and the extended DBGA for the stay-put probability $P(t)$, defined in Eq. (6.33) are compared in Fig. 6.2. Since, in the context of the double well problem, $P(t)$ measures the survival probability for the particle initially localized in the right well, $(1 - P(t))$ accounts for the probability of leakage across the barrier. It is interesting to note that there are many more coherent oscillations in the result for $P(t)$ in the extended DBGA than in the DBGA.

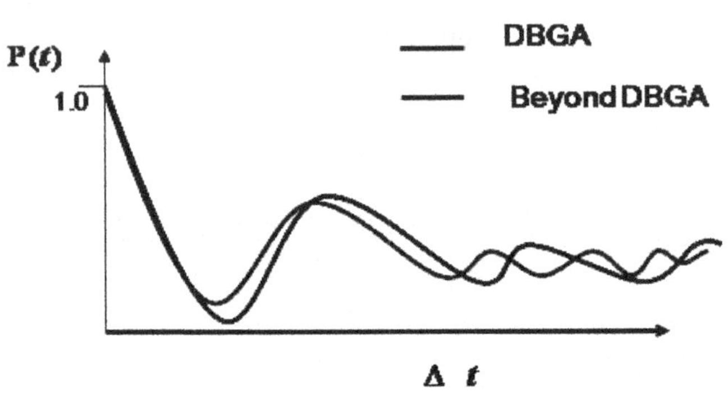

Fig. 6.2. The stay-put probability in DBGA (darker line) and beyond DBGA (lighter line). The later shows more oscillations, and hence enhanced coherence.

6.4. Dissipative Diamagnetism

(i) Preliminary Remarks

The problem of diamagnetism, or magnetism arising from the orbital current of electrons in a solid due to the Lorentz force of an applied magnetic field, has a long history in the subject of solid state physics. An isolated electronic orbit does of course create a magnetic moment in opposition to the applied field, as dictated by Lenz's law, but when a collection of such electrons is considered within a confined volume, statistical mechanics come into play. There exists a remarkable result due to van Leeuwen and Bohr that the classical partition function of such an ensemble of electrons is independent of the magnetic field. Because the diamagnetic moment is related to the logarithmic derivative of the partition function with respect to the field, the former as well as the associated susceptibility vanish. This conclusion goes under the name of Bohr-van Leeuwen theorem [2, 144].

What is not widely appreciated in the text book proofs of the Bohr-van Leeuwen theorem is the critical role of the boundary within which the electrons reside [2]. It turns out that while the bulk electrons do contribute to a finite diamagnetic moment, that contribution gets exactly nullified by the so-called 'skipping orbits' of the electrons which collide with the surface of the container and are bounced back (Fig. 6.3). This result is surprising at first sight because normally, the number of surface electrons is $\sim N^{2/3}$, and the ratio of this to the bulk electrons N, is $\sim N^{-1/3}$, which vanishes in the thermodynamic limit of statistical mechanics. But, the diamagnetic moment is proportional to the average value of the dynamical variable denoted by $< \vec{r} \times \vec{v} >$, where $\vec{r}$ is the position vector and $\vec{v}$ is the velocity vector of the electron. Therefore, when the origin of the co-ordinate system is chosen once and for all, a substantial number of electrons will have large $\vec{r}$-vectors even if their number is small. This is why the problem of diamagnetism has been classified by Peierls as one of the Surprises in Theoretical Physics, in that the boundary which is normally considered innocuous does play a crucial role [145].

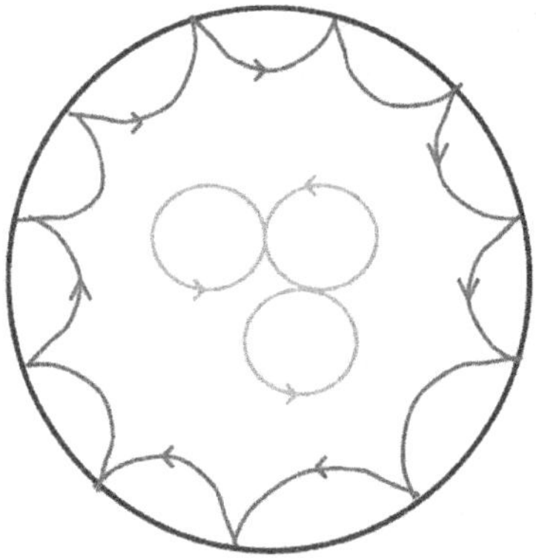

Fig. 6.3. The complete electron orbits in the bulk and the skipping orbits due to reflections at the boundary.

The conclusion of classical statistical mechanics about the absence of diamagnetism is clearly at variance with the reality of say, a metal like bismuth, which does exhibit a diamagnetic susceptibility. Indeed, diamagnetism may be viewed as one of the triumphs of quantum mechanics when Landau showed that the cancellation towards the diamagnetic moment because of bulk and surface electrons is incomplete, in quantum mechanics [146]. What was earlier a continuum of states gets split into equispaced Landau levels, mutually separated by $\hbar\omega_c$ where ω_c is the cyclotron frequency (Fig. 6.4). Summarizing then, we recognize that the phenomenon of diamagnetism is

(a) critically dependent on the boundary and,
(b) inherently and essentially quantal.

There is one other angle from which diamagnetism may be viewed to have contrasting perspectives. As mentioned before, the component of the magnetization along the z-axis in which the magnetic field H is applied, is given by

$$m_z = \frac{1}{\beta}\frac{\partial}{\partial H}\ln \mathcal{Z}, \tag{6.52}$$

where $\mathcal{Z}$ is the partition function. In the sense of Eq. (6.52), therefore, diamagnetism can be regarded as a themodynamic property, like the heat capacity, Pauli paramagnetism, etc, because $\ln \mathcal{Z}$ is related to the

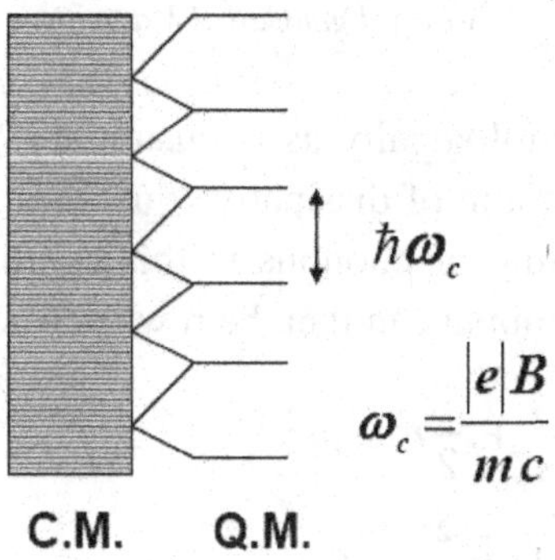

Fig. 6.4. The split Landau levels in quantum mechanics, in which the level separation is constant, given by the cyclotron frequency.

Helmholtz free energy. On the other hand, we have remarked above that m_z can also be expressed as

$$m_z = \frac{|e|}{2c} < (\vec{r} \times \vec{v})_z >, \tag{6.53}$$

where e is the electronic charge and c is the speed of light. Therefore a piece of the diamagnetic moment connotes also to the transport or the motion of the electron, much like the Drude conductivity, and hence it is not surprising that the methods of nonequilibrium statistical mechanics like the Langevin equation can also be employed in calculating m_z. As in the Drude problem the diamagnetic current can be evaluated from the dissipative equation of motion of the electron in the presence of the Lorentz force, and then m_z can be extracted from the asymptotic $t = \infty$ limit of the result [125, 133]. It is this twin property of diamagnetism, calculable on the one hand from the Gibbsian methods of equilibrium statistical mechanics, and derivable also from the Langevin equation which forms the Einstein strategy [147], allows it to be envisaged as the unifying thread between Gibbs and Einstein approaches to statistical mechanics.

(ii) Quantum Brownian Motion

Against the background of the preliminary remarks we want to set up the problem of Landau diamagnetism in the spirit of the Drude problem in which the electron is viewed to suffer collisions (or scattering) with other degrees of freedom such as phonons. But, instead of treating the collision parameters phenomenologically as is customary [148], we want to pose the issue within the realm of dissipative quantum mechanics in which the environment, consisting of phonons in this context, is treated explicitly. The complete Hamiltonian can then be rewritten as

$$\mathcal{H} = \frac{1}{2m}(\vec{p} - \frac{e}{c}\vec{A})^2 + \frac{1}{2}m\omega_0^2\vec{q}^2$$
$$+\sum_{j=1}^{N}\left[\frac{\vec{p}_j^{\,2}}{2m_j} + \frac{1}{2}m_j\omega_j^2\left(\vec{q}_j - C_j\frac{\vec{q}}{m_j\omega_j^2}\right)^2\right], \quad (6.54)$$

where m is the electron mass, $\vec{p}$ is the momentum operator, $\vec{A}$ is the vector potential $(\vec{H} = Curl\,\vec{A})$, $\vec{q}$ is the position coordinate operator, ω_o is the curvature of the confining harmonic oscillator potential, j refers to a given harmonic oscillator of the environment and C_j is a suitable coupling constant. In the spirit of the System-plus-Bath approach then

$$\mathcal{H}_s = \frac{1}{2m}(\vec{p} - \frac{e}{c}\vec{A})^2 + \frac{1}{2}\left[m\omega_o^2 + \sum_j\frac{C_j^2}{m_j\omega_j^2}\right]\vec{q}^2, \quad (6.55)$$

$$V = -\sum_{j=1}^{N}C_j\vec{q}.\vec{q}_j, \quad (6.56)$$

$$\mathcal{H}_B = \sum_{j=1}^{N}(\frac{\vec{p}_j^{\,2}}{2m_j} + \frac{1}{2}m_j\omega_j^2\vec{q}_j^{\,2}). \quad (6.57)$$

The confining potential (i.e. the second term on the right hand side of Eq. (6.54)) is a Darwin term [149], necessary for retrieving the correct boundary contributions, but to be set equal to zero at the end of the calculation.

It is important to point out that whereas in the System-plus-Bath approach the interaction term V is treated perturbatively, the linear coupling in Eq. (6.56), in both the system coordinate $\vec{q}$ and the bath coordinate $\vec{q}_j$, allows for an exact treatment of the Hamiltonian, as discussed below. Other than that however, the nature of the bath is identical to that in the spin-boson problem, as is clear from the second quantized form of Eq. (6.57). But, unlike in the master equation method, which amounts to the Schrödinger picture formulation of quantum mechanics, we present now an equation of motion treatment of the Hamiltonian in Eq. (6.54) as in the Heisenberg picture. The exact equation can be derived as [125, 133].

$$m\ddot{\vec{q}} + m\omega_0^2\vec{q} - \frac{e}{c}(\dot{\vec{q}} \times \vec{H}) + \int_0^t dt'\gamma(t-t')\,\dot{\vec{q}}(t') = \vec{F}(t). \tag{6.58}$$

Note that the Lorentz force (i.e. the third term on the left hand side of Eq. (6.58)) appears explicitly in a manner independent of the choice of any particular gauge for the vector potential $\vec{A}$. In Eq. (6.58) above,

$$\gamma(t) = \sum_{j=1}^{N} \frac{C_j^2}{m_j\omega_j^2}\cos(\omega_j t), \tag{6.59}$$

$$\vec{F}(t) = \sum_{j=1}^{N} C_j\left[\vec{q}_j(o) - C_j\frac{\vec{q}(o)}{m_j\omega_j^2}\right]cos(\omega_j t) + C_j\frac{\vec{p}_j(o)}{m_j\omega_j^2}\sin(\omega_j t). \tag{6.60}$$

It is interesting to observe that while Eq.(6.58) has the formal structure of a generalized Langevin equation with a memory (friction) kernel $\gamma(t)$, the only step towards achieving irreversibility we have taken is to provide a sense for the 'arrow of time' by integrating the equation of motion in the forward direction of time. The other point to note is that the 'noise' force $\vec{F}(t)$ lives in the Hilbert space or the phase space of the Hamiltonian $\mathcal{H}_B$. In order that the latter can be said to have the requisite properties of a heat bath we must go to the continuum limit in which the summation over j is replaced by an integral over the frequency of the bath excitations with the aid of a density of states $g(\omega)$ that is related to the spectral function $J(\omega)$introduced earlier (cf., Eq. (6.41). Thus

$$\sum_j \to N\int_0^\infty g(\omega)d\omega,\ m_j = m,\ C_j = \frac{C}{\sqrt{N}},$$

and

$$\gamma(t) = \frac{C^2}{m}\int_0^\infty d\omega \, \frac{g(\omega)}{\omega^2}\cos(\omega t). \qquad (6.61)$$

Therefore, the friction $\gamma(t)$ is quadratic in the coupling between the system and the bath, a consequence of the Gaussian nature of the bath, and akin to a response function that subsumes the entire spectrum of bath excitations. Further manifestation of the Langevin-like character of Eq.(6.58) ensues when we consider the spectral properties of the noise operator $\vec{F}(t)$, encapsulated by

$$< \{F_\alpha(t), F_\beta(t')\} >$$

$$= \delta_{\alpha\beta}\frac{2}{\pi}\int_0^\infty d\omega Re[\tilde{\gamma}(\omega + io^+)]$$

$$\hbar\omega\coth\left(\frac{1}{2}\beta\hbar\omega\right)\cos[\omega(t-t')]d\omega,$$

$$< [F_\alpha(t), F_\beta(t')] = \delta_{\alpha\beta}\frac{2}{i\pi}\int_0^\infty d\omega Re[\tilde{\gamma}(\omega + io^+)]\hbar\omega\sin[\,\omega(t-t')]d\omega\,,$$

(6.62)

where the Laplace transform of $\gamma(t)$ is defined by

$$\tilde{\gamma}\,(s) = \int_0^\infty dt\, e^{ist}\,\gamma(t)\,, Im\, s > 0. \qquad (6.63)$$

In Eq. (6.62), { , } represents a symmetric combination while [,] denotes a commutation. Further, the angular brackets represent a statistical average in the Gibbsian canonical ensemble in which the bath variables $\vec{q}_j$ and $\vec{p}_j$ are supposed to be equilibrated at the temperature T of the bath ($\beta = (k_B T)^{-1}$). The underlying philosophy of the quantum Langevin equation is then the following: The charged electron, starting from an initial position $\vec{q}(o)$ and initial (kinematic) momentum $m\dot{\vec{q}}\,(o)$, is subjected to a systematic (Lorentz) force due to an external magnetic field $\vec{H}$ and a dissipative force due to frictional coupling with a bath, kept at equilibrium at the temperature T.

As mentioned earlier, a commonly studied model for bath excitations is to assume Ohmic dissipation in which the density of states is taken to be of the form

$$g(w) = \frac{3\omega^2}{\omega_c^3}, \qquad \omega < \omega_c$$

$$= 0, \quad \omega > \omega_c, \tag{6.64}$$

where ω_c is a cut-off frequency, as before (cf., Eq. (6.42). For this case, the friction $\gamma(t)$ is given by (cf. Eq. (6.61))

$$\gamma(t) = \frac{3C^2}{m\omega_c^3} \frac{\sin(\omega_c t)}{t}. \tag{6.65}$$

If the cut-off frequency ω_c is extended to infinity the time-dependent function in Eq. (6.65) can be replaced by π times a delta function, thus

$$\gamma(t) = 2v_0 \, \delta(t), \tag{6.66}$$

where v_0 is a lumped parameter given by:

$$v_0 = \frac{3\pi C^2}{2m\omega_c^3}. \tag{6.67}$$

Under this condition Eq. (6.58) reduces to an ordinary Langevin equation (with memory-less friction):

$$m\ddot{\vec{q}} + m\omega_0^2\vec{q} - \frac{e}{c}(\dot{\vec{q}} \times \vec{H}) + 2v_0\dot{\vec{q}}(t) = \vec{F}(t), \tag{6.68}$$

though the noise-correlations continue to remain non-white, and hence non-Markovian, a signature of quantum fluctuations.

We may now proceed with the evaluation of the diamagnetic moment which from Eq. (6.53) can be rewritten as

$$m_z(t) = \frac{|e|}{2c} < x(t)\dot{y}(t) - y(t)\dot{x}(t) >. \tag{6.69}$$

The idea is to solve Eq. (6.68) explicitly with arbitrary boundary conditions, substitute the appropriate solutions in Eq. (6.69), calculate averages in accordance with the spectral properties of $\vec{F}(t)$ (cf. Eq. (6.62)), and extract $m_z(t)$ as a fully time-dependent quantity, akin to the Drude current. From the latter, we can take the asymptotic limit $t \to \infty$, when the electron is expected to thermalize with the bath and then set $\omega_o = 0$. The order in which these two limits are taken is crucial in order to capture the boundary contributions, as mentioned before. The result is:

$$m_z = \lim_{\omega_0 \to 0} \lim_{t \to \infty} m_z(t)$$
$$= -2k_B T (\frac{e}{mc})^2 \sum_{n=1}^{\infty} \frac{1}{(v_n + \gamma)^2 + \omega_c^2}, \tag{6.70}$$

where

$$\gamma = \frac{v_0}{m} \text{ and } v_n = 2k_B T\, n\pi, \tag{6.71}$$

and ω_c is the cyclotron frequency, as introduced earlier. Equation (6.70) then is the "Landau – Drude" expression for dissipative diamagnetism wherein γ has the interpretation of a scattering rate, *a la* Drude. Because the latter has its origin in the coupling with the heat bath we can retrieve the usual Landau result of equilibrium statistical mechanics by switching off this coupling (i.e. by setting $\gamma = 0$). The result is [133]

$$m_z^0 = \frac{|e|}{2mc}[\frac{2}{\beta\hbar\omega_c} - \coth(\frac{1}{2}\beta\hbar\omega_c)]. \tag{6.72}$$

Before concluding we may remark that the treatment presented above contains as a limiting case the quantum harmonic oscillator, a much-studied problem in dissipative quantum mechanics [129]. This can be easily seen from Eq. (6.54) by setting the vector potential $\vec{A} = 0$.

(iii) The Gibbs Approach:

As mentioned earlier, the process of extracting a thermal equilibrium property like the Landau diamagnetism in Eq. (6.70) from a time-dependent Brownian motion may be referred to as the Einstein approach to statistical mechanics [147]. Underneath is the assumption that as time t is allowed to increase to infinity all points in the phase space are explored and therefore, the system is ergodic. On the other hand, because the Hamiltonian in Eq. (6.54) is subjected to an exact treatment, its separation into the system part and the bath-part is not quite meaningful – the system and the bath are inexorably linked. From another point of view the full Hamiltonian in Eq. (6.54) can be viewed to encompass a whole many body system in its own right. The question then is: why not subject the Hamiltonian in Eq. (6.54) to the usual Gibbsian treatment of equilibrium statistical mechanics and derive the diamagnetic moment accordingly? The answer to this question is the subject of this subsection. If the consequent result matches with the one obtained from the Einstein approach, the concomitant analysis would unify these two distinct approaches to statistical mechanics. We may point out that such an equivalence between the Einstein and the Gibbs results is not readily available for a general quantum many body system.

While on the issue of a comparative analysis of the Einstein and the Gibbs approaches to statistical mechanics it is pertinent to underscore the different routes through which the temperature T enters into the discussion. In the Einstein approach the harmonic oscillator system described by the Hamiltonian in Eq. (6.57) is taken to be in thermal equilibrium at the temperature T. The system, viz. the charged quantum particle in a magnetic field, is viewed to be in an arbitrary nonequilibrium state but is eventually thought to come to equilibrium at the temperature of the harmonic oscillator system, through long-lasting interactions. In contrast, in the Gibbs approach, the entire system described by the Hamiltonian in Eq. (6.54) is assumed to be in thermal equilibrium with an invisible heat bath, not reckoned with in detail, at all. The entire system itself is a member of an ensemble of similar systems – its own replicas. The temperature T is defined by the energy fluctuations of the entire system governed by the Hamiltonian in Eq. (6.54) and the heat capacity [17].

The starting point of the Gibbs approach is the quantum partition function which is a functional integral over all relevant variables in terms of which we express the exponential of an Euclidean action $\mathcal{A}_e$:

$$\mathcal{Z} = \int D\,[\vec{x}]\ \exp(-\frac{1}{\hbar} A_e\,[\vec{x}]), \tag{6.73}$$

where

$$A_e = \int_0^\beta d\tau\, [\mathcal{L}_s(\tau) + \mathcal{L}_I(\tau) + \mathcal{L}_B(\tau)]. \tag{6.74}$$

The $\mathcal{L}'s$ represent the respective Lagrangians associated with the three terms in Eq. (6.55 - 6.57). The Helmholtz free energy is given by

$$F = -\frac{1}{\beta} ln\, \mathcal{Z}. \tag{6.75}$$

The equilibrium magnetization is now obtained from

$$m_z = \lim_{\omega_0 \to 0} -\frac{\partial F}{\partial H}, \tag{6.76}$$

where, it may be noted, we follow a similar procedure, viz., we take the derivative of the free energy calculated in the thermodynamic limit and then set the ('Darwinian') frequency ω_0 to zero. We quote the result for the general non-Ohmic dissipation:

$$m_z = -2k_B T\, H\, (\frac{e}{mc})^2 \sum_{n=1}^{\infty} \frac{1}{(v_n + \gamma(v_n))^2 + \omega_c^2}. \tag{6.77}$$

Clearly, the friction is now frequency dependent, $\gamma(v_n)$ being calculated from Eq. (6.63) by analytically continuing $\tilde{\gamma}(s)$ such that $Re\, s = v_n = 2k_B T\, n\pi$. The earlier result for Ohmic dissipation of course emerges when the frequency dependence of the friction is ignored.

One other quantify that can be easily calculated from the Gibbs formulation of statistical mechanics is the fluctuation property in equilibrium, for instance of the position, captured by the auto correlation function:

$$C(\tau) = \frac{1}{\mathcal{Z}} Tr\left[(\vec{q}(\tau).\vec{q}\,(o))e^{-\beta\mathcal{H}} \right]. \tag{6.78}$$

The correlation function can be extracted by adding to $\mathcal{H}$ a small energy due to an external force $\vec{f}$ and then employing the following functional derivatives:

$$< \vec{q}(\tau).\vec{q}(o) >= Tr\left(\frac{\delta}{\delta\vec{f}(\tau)}\frac{\delta}{\delta\vec{f}(o)}\frac{e^{-\beta\mathcal{H}}}{\mathcal{Z}}\right)_{\vec{f}=0}. \tag{6.79}$$

One of the remarkable results of statistical mechanics is that the spontaneous fluctuations described by the correlation function $C(\tau)$ can be related to the imaginary part of a frequency-dependent susceptibility which is the steady-state response of the time-dependent coordinate to the tiny force $\vec{f}$. This result of linear response theory can be expressed in terms of the Fluctuation – Dissipation Theorem:

$$\tilde{C}(w) = \frac{2\hbar}{1 - e^{-\beta\hbar\omega}} \chi''(\omega). \tag{6.80}$$

Once again, the equivalence between the Gibbs and the Einstein approaches can be established by calculating the susceptibility directly from the quantum Langevin equation [150]:

$$m\ddot{\vec{q}} - \frac{e}{c}(\dot{\vec{q}} \times \vec{H}) + \int_{-\infty}^{t} dt' \, \dot{\vec{q}}(t')\gamma(t - t') = \vec{F}(t) + \vec{f}(t), \tag{6.81}$$

and

$$< q_\gamma(t) > = \int_{-\infty}^{t} d\tau \, \chi_{\gamma\beta}(t - \tau)(f_\beta(\tau), \; \gamma, \beta = x, y, z. \tag{6.82}$$

Before concluding, we wish to stress that the dissipative Landau diamagnetism can indeed be viewed as a paradigm for coherence and decoherence phenomena. When the electron goes in a cyclotron orbit with a well defined frequency ω_c its quantum phase is preserved much like that of a tunneling particle, and hence, ω_c is akin to the tunneling frequency Δ. Therefore, the occurrence of Landau diamagnetism, which is a collective phenomenon that survives down to zero temperature – not considered here as it requires a treatment of the Fermionic wave function, is a signature of coherence. The effect of the environment, composed of bosonic-like excitations, is to introduce decoherence. Because decoherence makes a quantal system seemingly classical, the Bohr-van Leeuwen quenching of diamagnetism can be interpreted as an extreme decoherent behavior. With this is mind we reexamine the Landau–Drude result (cf., Eq. (6.70)) in terms of resistance, which is the most basic attribute of dissipation. We thus introduce a scaled resistance r that is the ratio of the Drude resistivity R_D to the Hall resistivity R_H:

$$r = \frac{R_D}{R_H}, R_D = \frac{m\gamma}{n_e e^2}, R_H = \frac{H}{n_e e c}, \tag{6.83}$$

where n_e is the number density of electrons. In terms of r, we have from Eq. (6.70).

$$-\frac{mc}{e} m_z = \frac{2k_B T}{\omega_c} \sum_{n=1}^{\infty} \frac{1}{(\bar{\mu}_n + r)^2 + 1}, \tag{6.84}$$

where

$$\bar{\mu}_n = \frac{v_n}{\omega_c}. \tag{6.85}$$

Note that $r = \frac{\gamma}{\omega_c}$ and therefore, r parametrizes the effective strength of dissipation vis-à-vis the 'coherence frequency' ω_c. In Fig. 6.5 we plot the left hand side of Eq. (6.84) versus r for two different values of the dimensionless parameter $\frac{2k_B T}{\hbar\omega_c}$. It is clear that larger values of r cause stronger decoherence. In addition, the persistence of coherence is more pronounced for larger cyclotron frequencies ω_c. Figure 6.5 is thus an illustration of how the system makes the transition from the coherent Landau regime to the decoherent Bohr-van Leeuwen regime, as the coupling to the environment, measured by r, is enhanced.

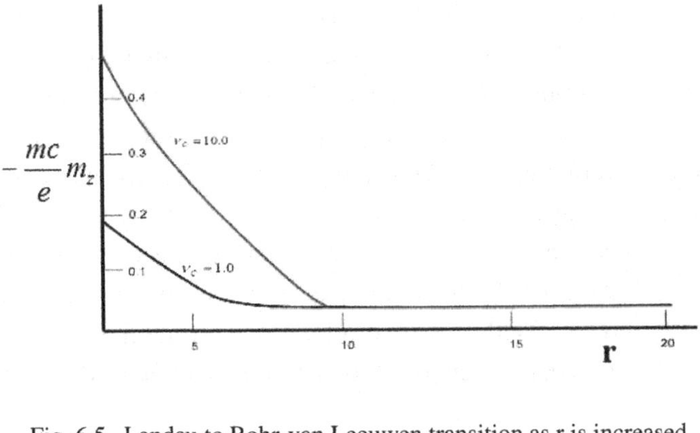

Fig. 6.5. Landau to Bohr-van Leeuwen transition as r is increased.

6.5. Spin Tunneling and Coherence-Decoherence Phenomenon

(i) Introductory Remarks

We have so far used two distinct magnetic paradigms for investigating dissipative quantum effects and associated coherence to decoherence transitions. In the spin-boson model the subsystem lives in a discrete (2 x 2) Hilbert space and coherence is characterized by quantum tunneling that corresponds to oscillatory behavior of the stay-put probability $P(t)$, as in Fig. 6.2. Decoherence is marked by an exponentially decaying $P(t)$, as a function of time. On the other hand, in the problem of dissipative diamagnetism, the Hilbert space of the subsystem is a continuous one, and coherence-to-decoherence transition is accompanied by the passage from the Landau regime to the Bohr-van Leeuwen regime. In both cases however decoherence effects are thought to be triggered by the coupling of the system to its environment, or enhanced thermal fluctuations due to increase of temperature. In the sequel we shall discuss yet another magnetic paradigm for which the Hilbert space of the system is still a discrete one but decoherence or classical-like effects are occasioned by simply increasing the size of the Hilbert space, but keeping fixed the coupling with the environment. The example in question concerns the topically important subject of spin tunneling in molecular magnets [134]. These are molecular clusters with complicated chemical formulae but can be referred to simply as Mn_{12} or Fe_8, for our purposes. Both these clusters have a large spin value $S = 10$ and have uniaxial anisotropy due to surrounding crystalline fields. As a result the basic spin Hamiltonian can be written as

$$\mathcal{H}_s = -DS_z^2, \tag{6.86}$$

where $D = 0.56k$ for Mn_{12} and $D = 0.275k$ for Fe_8. Thus there are 21 energy levels but barring the S = 0 level the other 20 levels are doubly-degenerate. The lowest energy is -100D corresponding to the eigenvalue of S_z being ± 10. A magnetic field, applied in the laboratory in a direction normal to the axis of quantization (z, in this case), can cause transitions amongst the 21 levels, which have been appropriately referred to in the literature as Spin Tunneling.

Note that if we add a constant term DS^2 to Eq. (6.86) and call $S_z = S\cos\theta$, as in the old semiclassical theory of quantum mechanics, then

$$\mathcal{H}_s = DS^2 sin^2\theta, \tag{6.87}$$

which has the same structure as the anisotropy energy of a single-domain nanomagnetic particle discussed in Chapter 5, if we identify the parameter DS^2 with kV (cf., Eq. (5.1). In the latter problem recall that the magnetization vector stochastically reorients itself amongst a continuum of values of θ, due to thermal fluctuations. In the molecular magnet described by Eq. (6.86), however, these transitions occur between a set of discrete levels that do not require thermal energy but can happen even at zero temperature because of quantum fluctuations. It is instructive to compare the relaxation time for a molecular magnet vis-à-vis a nanomagnetic particle, considered in Chapter 5 (See Fig. 6.6).

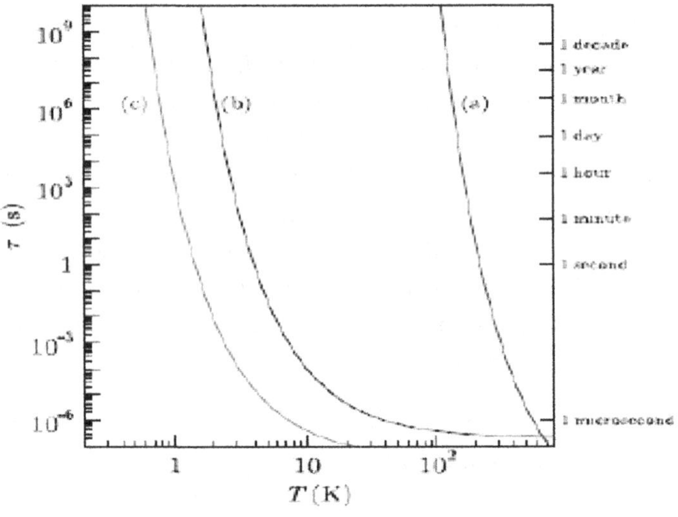

Fig. 6.6. The reorientation / tunneling times for nanomagnetic particles (a) and magnetic clusters of Mn_{12} (b) and Fe_8(c). Note the orders of magnitude difference in timescales.

In the light of the above remarks the problem posed by Eq. (6.86) is therefore truly the quantum cousin of the classical problem of super paramagnets, discussed earlier in Chapter 5. Hence, what we are discussing now is the effect of further size-reduction from nanomagnets to molecular clusters and how the dissipative behavior changes as a result of this miniaturization. It is also pertinent to remark that a molecular magnet of spin $S = 10$ is not a microscopic object but is described by an extended wave function covering a few angstroms and hence the associated spin tunneling is a quantum coherent phenomenon over mesoscopic length scales. What the environment does to this object in terms of decoherence is therefore a topic of great current interest in quantum information processes. Our discussion here is based on Ref. [135].

(ii) The Hamiltonian and Relaxation Dynamics

The Hamiltonian describing a large spin molecular magnet in interaction with its environment is indeed a generalized version of the spin-boson Hamiltonian (cf., Eq. (6.28)).

$$\mathcal{H} = \mathcal{H}_s + \sum_{\underline{q}} g_{\underline{q}} F(S) \left(b_{\underline{q}} + b_{\underline{q}}^{+} \right) + \mathcal{H}_B, \tag{6.88}$$

where $\mathcal{H}_s$ is the bath Hamiltonian given by Eq. (6.31) and $\mathcal{H}_s$ is the Hamiltonian of the system given by Eq. (6.86) (to which a term due to Zeoman coupling with an external magnetic field can be easily added, without much difficulty). The crucial difference with the spin-boson Hamiltonian however arises in the spin-dependent coupling term $F(s)$ which admits higher-order spin operators than were allowed in the earlier spin one-half case. We consider two distinct forms:

$$F(S) = \frac{1}{2} (\eta_+ S_- + \eta_- S_+), \tag{6.89}$$

which is akin to the case of the spin-boson Hamiltonian, and

$$F(S) = \eta_+ (S_z S_- + S_- S_z) + \eta_- (S_z S_+ + S_+ S_z), \tag{6.90}$$

which leads to 'nonlinear' spin coupling (that vanishes in the spin one-half limit). Here $\eta_\pm$ are (complex) constants.

Introducing the projection (or ladder) operator:

$$X_n^m = |n><m|, \tag{6.91}$$

where the eigenstates $|n>$ (or $|m>$) are defined by

$$S_z|n = n|n>, \tag{6.92}$$

and using the reduced density operator (cf., Eq. (6.18)) we may derive

$$\frac{d}{dt}X_n^m = i[\mathcal{H}_s, X_n^m] + R_n^m, \tag{6.93}$$

where the relaxation matrix R_n^m is

$$R_n^m = -\int_0^t d\tau\,(K(t-\tau)F(\tau)[F, X_n^m] - K(t-\tau)[F, X_n^m]F(\tau)), \tag{6.94}$$

$$K(\tau) = \int_0^\infty \frac{d\omega}{\omega} J(\omega)\left[n_\omega e^{i\omega\tau} + (n_\omega + 1)e^{-i\omega\tau}\right]. \tag{6.95}$$

The spectral density $J(\omega)$ has already been introduced (cf., Eq. 6.41) while n_ω are the Bose-Einstein factors (associated with the environmental bosons), defined by

$$n_\omega = \left[\exp\left(\hbar\omega/_{k_B T}\right) - 1\right]^{-1}. \tag{6.96}$$

As earlier, we restrict ourselves to the Born-Markov limit of the master equation (by extending the upper limit of the integral in Eq. (6.94) to $t = \infty$) and the so-called 'rotating wave approximation' [90] under which $X_n^m(t)$ can be replaced by

$$X_n^m(\tau) = e^{-i\Delta_{nm}(t-\tau)}X_n^m(t), \tag{6.97}$$

where

$$\Delta_{nm} = <n|\mathcal{H}_s|n> - <m|\mathcal{H}_s|m>. \tag{6.98}$$

(iii) The transverse susceptibility and results

The central quantity of interest in our attempt to describe the spin-relaxation behavior of molecular magnets is the transverse susceptibility that captures the frequency-dependent linear response to a weak oscillatory field applied along the x-axis. The latter, in linear response theory, is given by [87]:

$$\chi_{\perp}(\omega) = \frac{1}{i}\int_{0}^{\infty} dt e^{i\omega t} \langle [S_x(o), S_x(t)] \rangle_{av}, \tag{6.98}$$

where $S_x(t)$ is the usual Heisenberg time evolution of the operator $S_x(o)$ and $\langle ... \rangle_{av}$ denotes the statistical ensemble appropriate to the Hamiltonian in Eq. (6.88). Note that a quantity like $\chi_{\perp}(\omega)$ is probed in common resonance experiments [136]. Also, its classical counterpart is given by [87]

$$\chi_{\perp}^{cl}(\omega) = \beta\omega \int_{0}^{\infty} dt\, e^{i\omega t} [\langle S_x^2(o) \rangle_{av} - \langle S_x(o) S_x(t) \rangle_{av}]. \tag{6.100}$$

In presenting the results of computation for $\chi_{\perp}(\omega)$ we consider two different forms of the spectral density $J(\omega)$:

$$J(\omega) = \lambda\, \omega, \tag{6.101}$$

and

$$J(\omega) = \lambda\, \omega^3. \tag{6.102}$$

The first case of Eq. (6.101) corresponds to Ohmic damping considered earlier while the second case of Eq. (6.102) is relevant to acoustic phonons and yields 'super-ohmic' damping [151]. We also introduce a scaled parameter:

$$\sigma = \frac{DS^2}{k_B T}, \tag{6.103}$$

and normalize the susceptibility by dividing it by the Curie-form $\chi_0 = \frac{\hbar^2 S(S+1)}{k_B T}$. The imaginary part of the normalized susceptibility $\chi''(\omega)$ is plotted along the ordinate versus the frequency ω along the abscissa (Fig. 6.7). The vertical straight lines in Fig. 6.7 are the positions at which resonance transitions occur. Because the transition operator S_x connects only the adjacent eigenstates of S_z these resonance frequencies are:

$$\Delta_{m,m+1}\, m\pi = D(2m+1). \tag{6.104}$$

There are therefore 20 possible resonance lines but we focus in Fig. 6.7 only on the positive frequencies located at 0.5Δ, 1.5Δ, 2.5Δ, ..., 9.5Δ. If the lines are sharp (delta functions) we have complete quantum coherence. But the lines acquire widths in view of both finite temperature T and the strength ⋋ of the coupling to the environment (cf., Eq. (6.101)). Note that the intensities of the lines decrease as we move towards lower frequencies, the intensity almost vanishing for the line located at 0.5Δ, in view of the detailed balance factor $n_\omega e^{\hbar\omega/k_BT}$ (cf., Eq. (6.96)). In addition, there is an extra narrowing of the low-frequency peaks because the spin-phonon coupling $F(S) \sim S_z S_\pm$ yields an effective spin-dependent damping, proportional to ⋋ $(2m \pm 1)^2$.

We have stated before that the corresponding classical formula for the susceptibility is given by Eq. (6.100). The latter can be evaluated on the basis of the generalized Smoluchowski-Fokker-Planck equation, which is an extended version of the Fokker-Planck equation in (5.20), that is valid even for low-damping, in which regime inertial terms have to

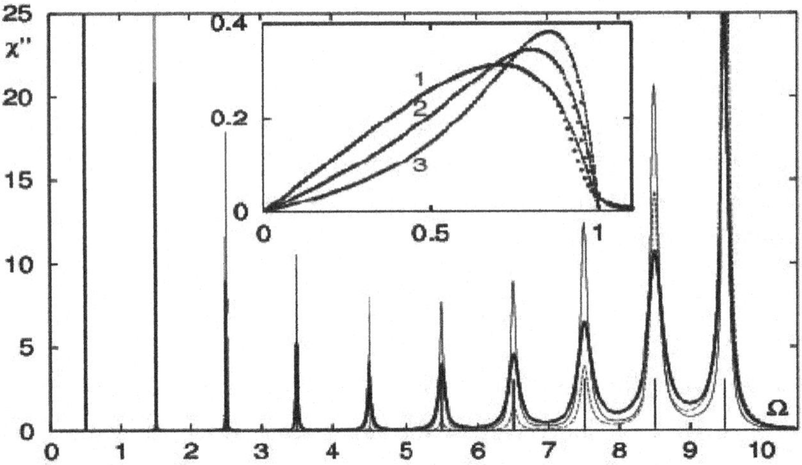

Fig. 6.7. The normalized transverse susceptibility $\chi_\perp(\omega)/\chi_0$ versus frequency ω for a spin S=10 with D=0.5k. Thick line: $\sigma = 5$ and ⋋ $= 3.10^{-8}$. Thin lines: effects of halving ⋋ with T fixed (solid) and halving T with ⋋ fixed (dashed). INSET: Exact classical susceptibility (cf., Eq. (6.105)) for zero dampling with $\sigma =$ 1,2 and 3 (lines) and Smoluchowski-Fokker-Planck result (dots) for ⋋ $= 0.003$.

be considered too [6.152]. It turns out that the zero-damping limit can be exactly calculated, and the result for the imaginary part of the susceptibility is [153, 154]

$$\chi''(\omega) = \frac{\pi\mu^2}{2k_BT}\frac{\hbar\omega}{Z}[1 - ({}^{\omega}/{\omega_\alpha})^2 \exp\left[\sigma({}^{\omega}/{\omega_\alpha})^2\right], \tag{6.105}$$

where μ is the size of the classical moment, Z is the partition function and ω_a is the frequency derived from the curvature of the classical potential (cf., Eq. (6.87)) at the two minima $\theta = 0$ and $\theta = \pi$. Note that the parameter σ (cf. Eq. (6.103) is also related to the barrier height renormalized by the thermal energy k_BT (cf., remarks following Eq. (6.87)). The expression given in Eq. (6.105) is plotted in the INSET of Fig. 6.7. It is interesting to point out that while the classical susceptibility is just a single-hump figure the corresponding quantum expression has a lot of structures comprising discrete spikes.

Our next item of attention is the quantum-classical crossover, first by keeping the environmental coupling λ fixed but increasing the value of S (Fig. 6.8: top), and second by keeping S fixed but increasing λ (Fig. 6.8:

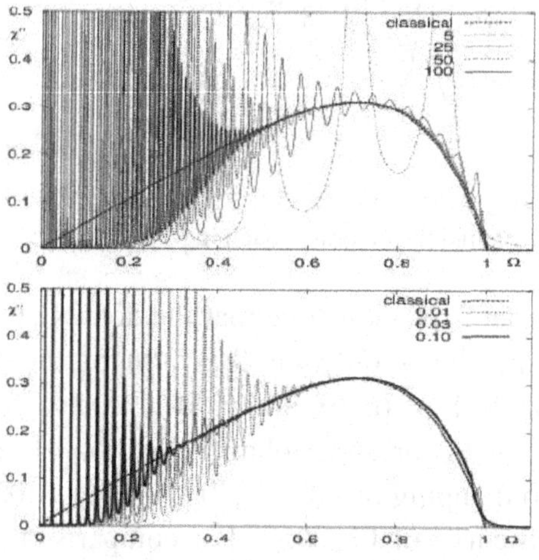

Fig. 6.8. The imaginary part of the susceptibility $\chi''(\omega)$ versus ω, for $\sigma = 1$. The thick dashed line is the classical result given by Eq. (6.105). Top: S = 5, 25, 50 and 100 with fixed ${}^{\lambda}/{}_{S} = 10^{-2}$. Bottom: fixed S = 50 with ${}^{\lambda}/{}_{S} = 10^{-2}$ (as in top), 3.10^{-2} and 10^{-1}.

bottom). The first case is interesting in its own right because, as S increases, the number of discrete energy levels increases too, but as the total energy is fixed, the levels become denser, ultimately approaching the continuum limit. Thus, for large S-values, the semi-classical approximation, captured by Eq. (6.87)), gains validity. It is intriguing to note that even an S-value of 100 mimicks the classical limit pretty well, even though smaller S-values are accompanied by large oscillations of $\chi''(\omega)$ - a signature of quantum coherence. The top of Fig. 6.8 thus encompasses the full range of behavior from quantum characteristics of molecular magnets to classical features of single-domain nanomagnetic particles, that we had discussed at length in Chapter 5.

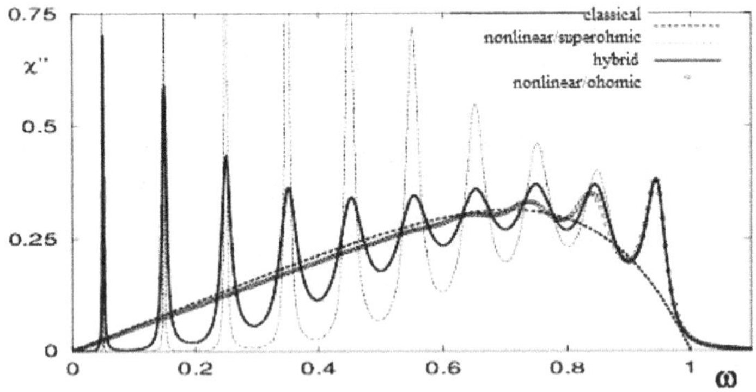

Fig. 6.9. $\chi''(\omega)$ versus ω for S=10 at $\sigma = 1$ with nonlinear spin coupling but both super-ohmic damping and ohmic damping.

Finally, we want to demonstrate that the details of quantum-classical cross-over do depend not only on the form of the coupling (i.e. the structure of $F(s)$, cf., Eqs. (6.89) and (6.90) but also on the nature of the spectral density: ohmic or super-ohmic (Eq. (6.101) and Eq. (6.102)). The classical zero damping result for $\chi''(\omega)$ (Eq. (6.105)) is given by the continuous dotted curve of Fig. 6.9. For comparison, we first show the results (dashes) for nonlinear spin coupling (cf. Eq. (6.90)): $F(s)\sim S_z S_\pm$, as well as super-ohmic damping (Eq. (6.102)). The second is a 'hybrid' case of still super-ohmic damping but linear spin coupling: $F(s)\sim S_\pm$, (cf., Eq. 6.89). The third graph is for nonlinear coupling $F(s)\sim S_z S_\pm$ but ohmic damping (circles). In order to make the comparison between the three quantum cases meaningful we have adjusted the λ value such that the intensity for the ground state transitions $\Delta_{9,10}$ (right most peaks) are all identical. We find that ohmic damping accentuates decoherence effects in that the oscillations are largely suppressed and 'classicality' is almost restored, even though the value of S is only moderately large ($S = 10$).

References

Chapter – 1

[1] W. Heisenberg, Z. Physik 49, 619 (1928); E.C. Stoner, Magnetism and Matter (Methuen and Co Ltd, London, 1934).
[2] J.H. Van Vleck, The Theory of Electric and Magnetic Susceptibilities. (Clarendon Press, Oxford, 1932).
[3] L.E. Bates, Modern Magnetism (Cambridge, 1951).
[4] D.C. Mattis, The Theory of Magnetism (Harper and Row, New York, 1965).
[5] G.T. Rado and H. Suhl, editors, Magnetism Vols. I – V (Academic Press, New York, 1973).
[6] P.W. Anderson, Basic Notions of Condensed Matter Physics (Benjamin, Menlo Park, CA, 1984).
[7] L.D. Landau and E.M. Lifshitz, Statistical Physics (Addison – Wesley, Reading, Mass, 1969).
[8] G. Parisi, Statistical Field Theory (Addison – Wesley, New York, 1988).
[9] S.K. Ma, Modern Theory of Critical Phenomena (Benjamin, Reading, Mass, 1976).
[10] E. Brezin, J.C. Le Guillou and J. Zinn-Justin, in Phase Transitions and Critical Phenomena, editors C. Domb and M.S. Green (Academic Press, New York, 1976).
[11] J. Zinn-Justin, Quantum Field Theory and Critical Phenomena (Clarendon Press, Oxford, 1989).
[12] N.D. Mermin and H. Wagner, Phys. Rev. Lett. 17, 1133 (1966).
[13] N.D. Mermin, Phys. Rev 176, 250 (1968).

[14] V.L. Berezinski, Ah. Eksp. Teor. Fiz. 59, 907 (1970) [Sov. Phys. JETP 32, 493 (1971)]

[15] Claude Itzykson and Jean-Michel Drouffe, Statistical Field Theory, Vol. 1 (Cambridge University Press, Cambridge, 1992).

[16] E. Ising, Z. Physik 31, 253 (1925).

[17] K. Huang, Statistical Mechanics, Second Edition (Wiley, New York, 1987).

[18] F. Keffer, Handbuch der Physik, Vol. XVIII (Springer-Verlag, Berlin, 1966).

[19] L. Onsager, Phys. Rev, 65, 117 (1944).

[20] See, for instance, K. Binder, in Materials Science and Technology Vol. 5: Phase Transformations of Materials, ed. By R.W. Cahn, P. Haasen and E.J. Kramer (VCH, Weinheim, 1991), p 405; and also, A.J. Bray, Adv. Phys. 43, 357 (1994).

[21] E. Salje, Phase Transitions in Ferroelastic and Co-elastic Crystals (Cambridge University Press, Cambridge, 1990).

[22] R.B. Potts, Proc. Camb. Phil. Soc. 48, 106 (1952).

[23] R.J. Birgeneau and P.M. Horn, Science 232, 329 (1986).

[24] H. Eugene Stanley, Introduction to Phase Transitions and Critical Phenomena (Oxford University Press, New York, 1987).

[25] K.G. Wilson and J.B. Kogut, Phys. Reports 12C, 75 (1974).

[26] Daniel J. Amit, Field Theory, the Renormalization Group and Critical Phenomena, revised second edition, (World Scientific, Singapore, 1984).

[27] Nigel Goldenfield, Lectures on Phase Transitions and the Renormalization Group (Addison-Wesley, Reading, Mass, 1982).

[28] R.B. Griffiths and J.C. Wheeler, Physics. Rev A2, 1047 (1970).

[29] See, for instance, H.B. Callen, Thermodynamics (John Wiley & Sons, Inc., New York, 1960).

[30] L.P. Kadanoff, Physics (NY) 2, 263 (1966).

[31] C. Domb and M.S. Green (eds.), Phase Transitions and Critical Phenomena, Vols. 5 and 6 (Academic Press, New York, 1976). See articles by A. Aharony; J. Als-Nielsen; E. Brezin, J. C. Le Guillon and J. Zinn-Justin; C. Di Castro and G. Jona-Lasino; L. Kadanoff; S.K. Ma; T.H. Niemeijer and J.M. Van Leeuwen; D. J. Wallace; and K.G. Wilson.

[32] A.A. Migdal, Z. Eksper, Teroet, Fiz. 69, 810, 1457 (1975) [Sov. Phys. JETP 42, 413, 743 (1975)]; L.P. Kadanoff, Ann. Phys. 100, 359 (1976).

[33] M. Blume, Phys. Rev. 141, 517 (1966); also in, Anharmonic Lattices, Structural Transitions and Melting, ed. T. Riste (Noordhoff, Leiden, 1974).

[34] H.W. Capel, Physica 32, 966 (1966).

[35] R.B. Griffiths, Phys. Rev. Lett. 24, 715 (1970).

[36] R.J. Birgeneau, G. Shirane, M. Blume and W.C. Koehler, Phys. Rev. Lett. 33, 1100 (1974).

[37] M. Blume, V.J. Emery and R.B. Griffiths, Phys. Rev. A4, 1071 (1971).

[38] D. Furman, S. Dattagupta and R.B. Griffiths, Phys. Rev. B15, 441 (1977).

[39] H. Haken, Rev. Mod. Phys. 47, 67 (1975) and in Synergetics (Springer, New York, 1977); A. Nitzan and J. Ross, J. Chem. Phys. 59, 241 (1973); G. Nicolis and J.W. Turner, Ann. N.Y. Acad. Sci. 316, 251 (1979).

[40] G.S. Agarwal and S. Dattagupta, Phys. Rev. A 26, 880 (1982); the reference to similar multicritical phenomena, seen in anisotropic antiferromagnets under external magnetic fields, can be found in M.E. Fisher and D.R. Nelson, Phys. Rev. Lett. 32, 1950 (1974) and D.R. Nelson and E. Domany, PRB 13, 236 (107).

Chapter – 2

[41] L.D. Landau, Phys. Z. Sowejetunion 11, 26 (1937); reprinted in Collected Papers of L. D. Landau, ed. D. ter Haar (Pergamon, New York, 1965), and V.L. Ginzburg, Fiz. Trend. Tela. 2, 2031 (1960) [Sov. Phys. Solid State 2, 1824 (1961)].

[42] B.K. Chakrabarti, A. Dutta and P. Sen, Quantum Ising Phases and Transitions in Transverse Ising Models, Lecture Notes in Physics 41 (Springer-Verlag, Heidelberg, 1996).

[43] S. Sachdev, Quantum Phase Transitions (Cambridge University Press, Cambridge, 1999).

[44] See, for instance, S.L. Sondhi, S.M. Girvin, J.P. Carini and D. Shahar, Rev. Mod. Phys. 69, 315 (1997).
[45] C. Kittel, Introduction to Solid State Physics (John Wiley, New York, 1971).
[46] J.M. Luttinger and L. Tisza, Phys. Rev. 70, 954 (1946).
[47] D. Bitko, T.F. Rosenbaum and G. Aeppli, Phys. Rev. Lett. 77, 940 (1996).
[48] D. Bitko, Ph.D. Thesis, University of Chicago, Illinois (1992, unpublished).
[49] M. Suzuki, Prof. Theo. Phys. 56, 1454 (1976).
[50] M. Barma and B.S. Shastry, Phys. Rev B 18, 3351 (1978).
[51] A. Aharony and B.I. Halperin, Phys. Rev B 18, 3351 (1978)
[52] V. Banerjee and S. Dattagupta, Phys. Rev. B64, 024427 (2001).
[53] R.P. Feynmann, Statistical Mechanics (W.A. Benjamin, Reading, Mass, 1972).
[54] H.F. Trotter, Proc. Amer. Math. Sec. 10, 545 (1959).

Chapter – 3

[55] C.A. Angell, Science 267, 1924 (1995).
[56] P. De Benedetti, Metastable Liquids (Princeton University Press, Princeton, New Jersey, 1997).
[57] C.A. Angell, J. Phys. Chem. Solids 49, 863 (1988).
[58] H. Vogel, Z. Phys. 22, 645 (1921); G.S. Fulcher, J. Am. Ceram. Soc. 8, 339 (1925).
[59] W. Kauzmann, Chem. Rev. 48, 219 (1948).
[60] G. Adams and J.H. Gibbs, J. Chem. Phys. 43, 139 (1965).
[61] T.R. Kirkpatrick and P.G. Wolynes, Phys. Rev. A. 35, 3072 (1987); Phys. Rev. B 36, 8552 (1987).
[62] T.R. Kirkpatrick, D. Thirumalai and P.G. Wolynes, Phys. Rev. A. 40, 1045 (1989).
[63] R. Monasson, Phys. Rev. Lett. 75, 2847 (1995).
[64] G. Parisi, in Complex Behaviour of Glassy Systems: Proceedings of the XIV Sitges Conference, ed. M. Rubi and C. Perez-Vicente (Springer, Berlin, 1997).

[65] G. Parisi, J. Phys. A. 30, L 765 (1997) Phys. Rev. Lett. 79, 3660 (1997).
[66] W. Wu, B. Ellman, T.F. Rosenbaum, G. Aeppli and D.H. Reich, Phys. Rev. Lett. 67, 2076 (1991); W. Wu, D. Bitko, T.F. Rosenbaum and G. Aeppli, 71, 1919 (1993).
[67] J. Mydosh, J. Mag. and Mag. Materials 7, 237 (1978); K. Binder, Festkörperprobleme XVII, 55 (1977).
[68] E. Courtens, Phys. Rev. Lett. 52, 69 (1984), and Phys. Rev. B 33, 2975 (1986).
[69] D. Sherrington and S. Kirkpatrick Phys. Rev. Lett. 35, 1792 (1975).
[70] S. Kirkpatrick and D. Sherrington, Phys. Rev. B17, 4384 (1978).
[71] S.F. Edwards and P.W. Anderson, J. Phys. F5, 965 (1975).
[72] K.H. Fischer and J.A. Hertz, Spin Glasses (Cambridge University Press, Cambridge, 1991).
[73] J.R. de Almeida and D.J. Thouless, J. Phys. A11, 983 (1978).
[74] T.K. Kopec, Phys. Rev. B 54, 3367 (1996).
[75] Non-Debye Relaxation in Condensed Matter, ed. By T.V. Ramakrishnan and M. Rajlakshmi (World Scientific, Singapore 1987).
[76] See, for instance, S. Dattagupta, Hyperfine Interactions 49, 253 (1989).
[77] S.F. Edwards and P.W. Anderson, J. Phys. F6, 1927 (1976).
[78] J.L. Lebowitz and O. Penrose, Physics Today., 23 (1973); also see R. Balescu, Equilibrium and Nonequilibrium Statistical Mechanics (John Wiley, New York, 1975).
[79] V. Banerjee and S. Dattagupta, Phys. Rev. B 68, 4202 (2003); also, Phase Transitions 77, 525 (2004).
[80] Z. Kutnjak, R. Pirc, A. Levstik, I. Levstik, C. Filipie, R. Blinc and R. Kind, Phys. Rev. B 50, 12421 (1994).
[81] See, for instance, S. Dattagupta and L.A. Turski, Phys. Rev. E 47, 1222 (1993).
[82] For reviews of mode coupling theories, see W. Götze, in Liquids, freezing and the glass transition, ed. By D. Levesque, J.P. Hansen and J. Zinn-Justin (Elsevier, New York, 1991); S.P. Das, Rev. Mod. Phys. 76, 785 (2004).

[83] A comprehensive review can be found in C. Dasgupta, in Proceedings of the 22nd IUPAP International Conference on Statistical Physics ed. S. Dattagupta et al (Indian Academy of Sciences, Bangalore 2004).

[84] G.I. Menon and C. Dasgupta, Phys. Rev. Lett. 73, 1023 (1994).

[85] M. Mezard and G. Parisi, Phys. Rev. Lett. 82, 747 (1999); J. Chem. Phys. 111, 1076 (1999).

Chapter – 4

[86] A recent textbook is: V. Balakrishnan, Elements of Nonequilibrium Statistical Mechanics (Ane Books India, New Delhi, 2008).

[87] See for instance, S. Dattagupta, Relaxation Phenomena in Condensed Matter Physics (Academic Press, Orlando 1987).

[88] S. Dattagupta and S. Puri, Dissipative Phenomena in Condensed Matter (Springer – Verlag, Heidelberg, 2004).

[89] A. Abragam, The Theory of Nuclear Magnetism (Oxford University Press, London 1961).

[90] See, for instance, G.S. Agarwal, Quantum Optics, Springer Tracts in Modern Physics 70, ed. By G. Hohler (Springer-Verlag, Berlin, 1974).

[91] R. Kubo, in Fluctuation, Relaxation and Resonance in Magnetic Systems, ed. By D. Ter Haar (Oliver and Boyd, Edinburgh 1962), p.23.

[92] See, for instance, N.W. Ashcroft and N.D. Mermin, Solid State Physics (Holt, Rinehart and Winston, New York, 1976), Chap.1.

[93] P. Debye, Polar Molecules (Dover, New York, 1929).

[94] A.S. Nowick and B.S. Berry, Anelastic Relaxation in Crystalline Solids (Academic Press, London, 1972).

[95] D. Kumar and S. Dattagupta, J. Phys C 16, 3779 (1983).

[96] R.J. Glauber, J. Math. Phys. 4, 294 (1963).

[97] M. Suzuki and R. Kubo, J. Phys. Soc. Jpn. 24, 51 (1968).

[98] K. Kawasaki, Phys Rev. 145, 224 (1966); also in Phase Transitions and Critical Phenomena Vol. 2, ed. By V. Domb and M.S. Green (Academic Press, London, 1972), P.443.

[99] P.W. Anderson and P.R. Weiss, Rev. Mod. Phys. 25, 269 (1953).
[100] N. Bloembergen, E.M. Purcell and R.V. Pound, Phys. Rev. 73, 679 (1948).
[101] K. Binder, Z. Phys. 267, 313 (1974), and Phys. Rev. B 15, 4425 (1977).
[102] J.W. Cahn and J.E. Hilliard, J. Chem. Phys. 28, 258 (1958); J. Chem. Phys. 31, 688 (1959)
[103] I.M. Lifshitz and V.V. Slyzov, J. Phys. Chem. Solids 19, 35 (1961).
[104] D. A. Huse, Phys. Rev. B 34, 7845 (1986).

Chapter – 5

[105] J. Frankel and J. Dorfman, Nature, 126, 274 (1930); C. Kittel, Phys. Rev. 70, 965 (1946).
[106] L. Neel, Ann. Geophys. 5, 99 (1949).
[107] C.P. Bean and J.D. Livingstone, J. Appl. Phys. 30, 1205 (1259); I.S. Jacobs and C.P. Bean, in Magnetism, ed. By G.T. Rado and H. Suhl (Academic, New York, 1953), Vol. III.
[108] R.E. Rosensweig, Ferrohydrodynamics (Dover, New York, 1997).
[109] H. Brooks, Phys. Rev. 58, 909 (1940).
[110] P. Bruno. Phys. Rev. B 39, 865 (1989).
[111] A. Einstein, Ann Physik 17, 549 (1905); Ann. Physik 19, 371 (1906).
[112] W.F. Brown, Jr., J. Appl. Phys. 30, 130 (1959); Phys. Rev. 130, 1677 (1963).
[113] H. Kramers, Physica 7, 284 (1940); also, S. Chandrasekhar, Rev. Med. Phys. 15, 1 (1943).
[114] G.S. Agarwal, S. Dattagupta and K.P.N. Murthy, J. Phys. C 17, 6869 (1984).
[115] See for instance, F. Reif, Fundamentals of Statistical and Thermal Physics (Mc Graw Hill, New York, 1965).
[116] P. Street and J. C. Woolley, Proc. Phys. Soc. London Sect. A 62, 562 (1949); E.P. Wohlfarth, J. Phys. F 10, L 241 (1980).

[117] D.C. Douglass, J.P. Bucher and L.A. Bloomfield, Phys. Rev. Lett. 68, 1774 (1992); D.C. Douglass, A.J. Cox, J.P. Bucher and L.A. Bloomfield, Phys. Rev. B 47, 12874 (1993); A.J. Cox, D.C. Douglass, J .G. Louderback, A.M. Spencer and L.A. Bloomfield, Z. Phys. D 26, 319 (1993).

[118] S. Dattagupta and S.D. Mahanti, Phys. Rev. B 57, 10244 (1998).

[119] R.K. Das, A. Konar and S. Dattagupta, Phys. Rev. B 71, 014442 (2005).

[120] S. Chakraverty, M. Bandyopadhyay, S. Chatterjee, S. Dattagupta, A. Frydman, S. Sengupta and P.A. Sreeram, Phys. Rev. B 71, 054401 (2005); also see M. Bandyopadhyay and S. Dattagupta, Phys. Rev. B 74, 214410 (2006).

[121] M. Blume, in Hyperfine Structure and Nuclear Radiations, ed. By E. Mathiah and D.A. Shirley (North – Hollard, Amsterdam, 1968).

[122] S. Dattagupta and M. Blume, Phys. Rev. B 10, 4540 (1974).

Chapter – 6

[123] S. Datta, Electronic Transport in Mesoscopic Systems (Cambridge University Press, Cambridge, 1995).

[124] Y. Imry, Introduction to Mesoscopic Physics (Oxford University Press, London, 1997).

[125] G.W. Ford, J.T. Lewis and R.F.O's Connell, Phys. Rev. A 37, 4419 (1988).

[126] F. Bloch, Phys. Rev. 105, 1206 (1957).

[127] A. G. Redfield, IBM J. Research Development 1, 19 (1957); Adv. Magn. Reson 1, 1(1965).

[128] S. Chakravarty, Phys. Rev. Lett. 49, 681 (1982); A.J. Bray and M.A. Moore, Phys. Rev. Lett. 49, 1545 (1982).

[129] U. Weiss, Quantum Dissipative Systems (World Scientific, Singapore, 1993).

[130] S. Chakravarty and A.J. Leggett, Phys. Rev. Lett. 52, 5 (1984).

[131] A. Garg, Phys. Rev. Lett. 77, 964 (1996).

[132] J. Leggett, S. Chakravarty, A.T. Dorsey, M.P.A. Fisher, A. Garg. and W. Zwerger, Rev. Mod. Phys. 59, 1 (1987).

[133] S. Dattagupta and J. Singh, Pramana 47, 211 (1996) and Phys. Rev. Lett. 79, 961 (1997).

[134] S.J. Blundell and F.L. Pratt, J. Phys. Cond. Matt. 16, R 771 (2004).

[135] J.L. Garcia-Palacios and S. Dattagupta, Phys. Rev. Lett. 95, 190 401 (2005).

[136] See [89] and C.P. Slichter, Principles of Magnetic Resonance (Harper and Row, New York, 1963).

[137] Hydrogen in Metals Vol. I, Topics in Applied Physics 28, ed. By G. Alefeld and J. Völkl (Springer-Verlag, Berlin, 1978).

[138] T. Holstein, Ann. Phys. (N.Y.) 8, 325 (1959); 8, 343 (1959); I.J. Lang and Y.A. Firsov, Sov. Phys. JETP 16, 1301 (1962); Y.A. Firsov, Polarons (Nauka, Moscow, 1975).

[139] L.D. Chang and S. Chakravarty, Phys. Rev. B 31, 154 (1985).

[140] M. Simonius, Phys. Rev. Lett. 40, 980 (1978).

[141] M. Sanjay Kumar, S. Dattagupta and N. Kumar, Phys. Rev. B 65, 134 501 (2002).

[142] U. Weiss and M. Wollensak, Phys. Rev. Lett. 62, 1663 (1989).

[143] T. Qureshi, Phys. Rev. B 53, 3183 (1996).

[144] N. Bohr, Ph.D. Thesis 1911, in Collected Works Vol. 1 (North-Holland, Amsterdam, 1972); H. J. van Leeuwen, J. Phys. (Paris) 2, 361 (1921).

[145] R. Peierls, Surprises in Theoretical Physics (Princeton University Press, Princeton, 1979).

[146] L. Landau, Z. Phys. 64, 629 (1930).

[147] L.P. Kadanoff, Statistical Physics – Statics, Dynamics and Renormalization (World Scientific, Singapore, 2000).

[148] C.W.J. Beenakker and H. van Houten, in Solid State Physics, 44, eds. H. Ehrenreich and D. Turnbull (Academic Press, New York, 1991).

[149] G. Darwin, Proc. Cambridge Philos. Soc. 27, 86 (1930).

[150] M. Bandyopadhyay and S. Dattagupta, Jour. Stat. Phys. 123, 1273 (2006) and J. Phys. Cond. Matt. 18, 10029 (2006).

[151] See for example, H. Grabert and H.R. Schober, in Hydrogen in Metals Vol. II, ed. By H. Wipf (Springer – Verlag, Heidelberg, 1994).

[152] See for instance, N.G. Van Kampen, Stochastic Processes in Physics and Chemistry. (North-Holland, Amsterdam, 1981).

[153] R.S. Gekht, Fiz. Met. Metalloved 55, 225 (1983) [Phys Met Metallogr 55, 12 (1983)].

[154] D.A. Garanin, V.V. Ischenko and L.V. Panina, Theor. Math. Phys. (Engl. Transl.) 82, 169 (1990).

Index